Understand the Weather

Peter Inness

For UK order enquiries: please contact Bookpoint Ltd,
130 Milton Park, Abingdon, Oxon OX14 4SB.
Telephone: +44 (0) 1235 827720. Fax: +44 (0) 1235 400454.
Lines are open 09.00–17.00, Monday to Saturday, with a 24-hour
message answering service. Details about our titles and how to
order are available at www.teachyourself.com

For USA order enquiries: please contact McGraw-Hill
Customer Services, PO Box 545, Blacklick, OH 43004-0545, USA.
Telephone: 1-800-722-4726. Fax: 1-614-755-5645.

For Canada order enquiries: please contact McGraw-Hill Ryerson
Ltd, 300 Water St, Whitby, Ontario L1N 9B6, Canada. Telephone:
905 430 5000. Fax: 905 430 5020.

Long renowned as the authoritative source for self-guided
learning – with more than 50 million copies sold worldwide –
the **Teach Yourself** series includes over 500 titles in the fields of
languages, crafts, hobbies, business, computing and education.

British Library Cataloguing in Publication Data: a catalogue record
for this title is available from the British Library.

Library of Congress Catalog Card Number: on file.

First published in UK 2008 by Hodder Education, part of Hachette UK,
338 Euston Road, London NW1 3BH.

First published in US 2008 by The McGraw-Hill Companies, Inc.

This edition published 2010.

Previously published as Teach Yourself Weather

The **Teach Yourself** name is a registered trade mark of
Hodder Headline.

Typeset by MPS Limited, a Macmillan Company.

Printed in Great Britain for Hodder Education, an Hachette UK
Company, 338 Euston Road, London NW1 3BH, by CPI Cox &
Wyman, Reading, Berkshire RG1 8EX.

The publisher has used its best endeavours to ensure that the URLs
for external websites referred to in this book are correct and active
at the time of going to press. However, the publisher and the author
have no responsibility for the websites and can make no guarantee
that a site will remain live or that the content will remain relevant,
decent or appropriate.

Hachette UK's policy is to use papers that are natural, renewable
and recyclable products and made from wood grown in sustainable
forests. The logging and manufacturing processes are expected to
conform to the environmental regulations of the country of origin.

Impression number	10 9 8 7 6 5 4 3 2 1
Year	2014 2013 2012 2011 2010

Acknowledgements

First of all, thanks to Ross Reynolds for encouraging me to try writing a book in the first place. Thank you also to all my colleagues at Reading for providing an atmosphere in which to work where there is such a wide diversity of meteorological knowledge, making it easy to find out things I didn't know. The same goes for former colleagues at the Met Office College and the Hadley Centre. Most of all thanks to my wife for all the support over the months of writing in the evenings. Finally thanks to my son Tom for lending me his drawing stuff.

The author and publisher would like to thank the following for permission to use their photographs and data:

Colour plates

© M. Ambaum, Plate 14
© Alec Bennett, Plates 15 and 23
© Neil Blanchonnet, Plate 7
© Laura Wilcox, Plate 27
© Peter Inness, Plates 16, 17, 24, 25
© Intergovernmental Panel on Climate Change 2007 report, 'Climate Change: The Physical Basis', Plate 26
© Tom Lyons, Murdoch University, Australia, Plate 8
© Met Office, Plate 19
©NASA Earth Observatory (earthobservatory.nasa.gov), Plates 5 and 13
© NERC Satellite Receiving Station, University of Dundee, Plates 1, 2, 3, 4, 6, 9, 10, 11, 12, 20, 21, 22
© I. Ptashnik, Plate 18

Black and white figures
Figures 3.8, 3.9a, 3.9b, 3.10a, 3.10b, 3.11a, 3.11b, 10.1, based on data from NOAA Earth Systems Research Laboratory (www.cdc.noaa.gov).

Figure 6.2, based on data from NOAA National Hurricane Center (www.nhc.noaa.gov).
Figure 6.5, based on data from NOAA Storm Prediction Center.
Figures 8.1 and 8.5 © Met Office.
Figure 8.2, based on a figure from J. Bjerknes and H. Solberg, 'Life cycle of cyclones and the polar front theory of atmospheric circulation', Geofisiske Publikationer, 1922.
Figures 9.1 and 9.3, based on data from Met Office.
Figure 10.2, based on data from Met Office Hadley Centre.

Credits

Front cover: © Rosemary Calvert/Alamy

Back cover: © Jakub Semeniuk/iStockphoto.com, © Royalty-Free/Corbis, © agencyby/iStockphoto.com, © Andy Cook/iStockphoto.com, © Christopher Ewing/iStockphoto.com, © zebicho – Fotolia.com, © Geoffrey Holman/iStockphoto.com, © Photodisc/Getty Images, © James C. Pruitt/iStockphoto.com, © Mohamed Saber – Fotolia.com

Contents

Foreword

As time goes on so the weather hits the headlines more and more. There is no doubt that the climate is changing and that human activities have caused it. I frequently give lectures on global warming under the title 'The Ultimate Weapon of Mass Destruction' and, if this book gives people an understanding of things meteorological and climatological, then as a certain British supermarket would say, 'every little helps'.

I am often asked how I became interested in the weather. I am not really sure, as it happened at a very early age. I think it was probably due to my parents' tales of the terrible winter of 1947, together with my own experience of the coverage in 1952 of the 'Great Smog' in London and the Lynmouth, Devon flood, followed shortly after by the disastrous North Sea floods of 1953. About the same time I started my own weather station, enlisting some of my friends, and went on to engineer my schooling so as to ensure I got the correct mathematics and physics qualifications to enter the Met Office. I also read, learnt and inwardly digested a copy of the first *Teach Yourself Meteorology* and I used the facts I had gleaned from it in my job interview with the Met Office. As a result I passed and joined in 1962 and stayed for 42 years.

I never wanted to be a TV weatherman and if I had known what would happen in 1987 I would never have allowed myself to be 'volunteered'. If only I had a penny for each time the infamous 'Hurricane' clip has been broadcast I would be a multi-millionaire. As it is I think I will have it engraved on my tombstone, when the time comes! It is even more unfortunate as my remark had nothing to do with the storm; I was not even on duty the afternoon and evening before – that's the press for you!

I feel I must mention the founding father of the Met Office, Admiral FitzRoy. He suffered from the frustrations of Civil Service bureaucracy, rivalry and workplace jealousy, which eventually led to his suicide in 1865. He never had the slightest inkling of global warming and he would be amazed at the meteorological advances of the last few years. A wrong forecast is now a rarity, whereas in his day a correct one was virtually unheard of. One hardly dares to predict what will happen in the next 150 years. Fittingly he left us with the words 'the problem with the weather is that the only certain thing about it is its unpredictability.' Something this book will help to dispel.

The book is invaluable for all interested in the weather. It takes you through a description of the atmosphere and how it works, to the weather systems themselves and then onto how an individual can produce his or her own forecast. It also takes you into the inner workings of a Met Office, how the data is collected and then used by giant computers to forecast the weather up to 16 days ahead. It details local weather and wild weather (including giving detail on my storm!) and finally gives an all important background to global warming and what, if anything, we can do to stop it.

This is a fascinating book, made better by a reference to me on page 114! I only hope that many of you will read it, as I did its predecessor 45 years ago, and end up, like me, with a long and fascinating career in the Met Office.

Michael Fish MBE DSc FRMetS, 2008

Meet the author

Welcome to *Understand the Weather*!

The weather provides us with some of the most dramatic natural spectacles on Earth. Hurricanes, tornadoes, electrical storms and flash floods all make fascinating viewing and trying to better understand and predict such phenomena is the reason many professional meteorologists began their careers. For me however it was the subtleties involved in the weather that got me interested in the subject. Why does the weather vary over quite short distances? When in the mountains, why does one valley experience very different conditions from the next valley only a few miles away over the ridge? Is it possible to predict, on a warm summer day, where any showers will occur by studying the local terrain and considering the direction of the winds? It is possible to start answering these questions without needing any specialized, expensive equipment. Simply observing weather patterns over a period of time, using our eyes to look at the development of clouds, feeling the direction of the winds on our face and sensing changes in temperature with our bodies allows us to start the process of understanding why the weather happens in the way that it does. We can start to see repeating patterns but we also begin to realize that the weather never goes through *exactly* the same patterns twice.

The fact that the weather goes through repeating patterns or cycles allows us to start making our own forecasts, without the need for complicated mathematical models and expensive computers. But it is the subtle variations in each individual day's weather that makes forecasting such a tricky problem and means that we're unlikely to be able to beat the professionals with all their sophisticated tools. It's good fun to try though, and with the increasing availability of weather information through the internet we can even get our hands on some of the tools that the professionals use to make their forecasts.

In this book we'll take a look at how the atmosphere works, from the large-scale circulations of winds around the globe down to the very small-scale effects that cause local variations in weather conditions. We'll look at how the professionals make their forecasts and think about how we might be able to make our own. Even when I'm away from all the sophisticated tools that I have available as a professional meteorologist I still get a lot of enjoyment from trying to predict the weather simply by watching the sky and noting changes in clouds and winds. Hopefully using this book will help you to do the same and experience the satisfaction of seeing your own forecasts turn out correct.

Peter Inness, 2010

Only got a minute?

The Earth's atmosphere is an extremely complex system. The weather patterns that we see on a daily basis are the result of the interaction between many different laws of physics. To fully understand how the atmosphere works takes many years of scientific study. That said, it is still possible to learn a lot about the weather simply by watching it unfold and observing its subtle patterns and trends. The weather at any given point on the Earth's surface is the result of very large-scale patterns of wind as the atmosphere works to move heat from the tropical regions towards the poles. These patterns are then modified by the surface over which they are passing which creates the variations in weather from place to place that we constantly observe. Even without a higher degree in atmospheric physics we can start to understand these processes and use our knowledge to predict how the weather will change in the future.

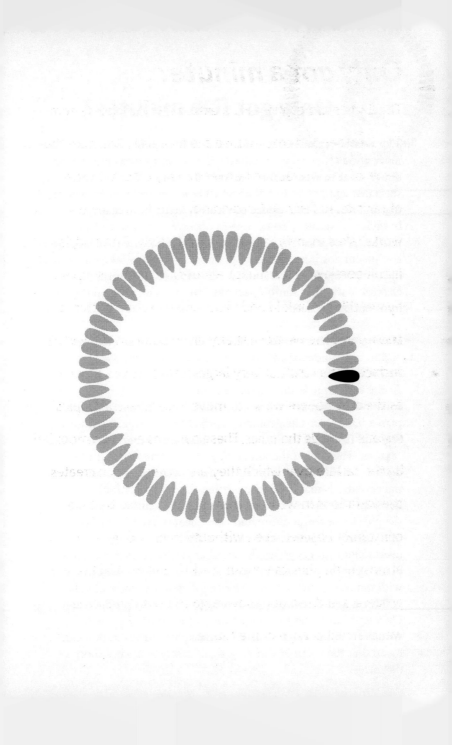

5 Only got five minutes?

The weather is an extremely accessible branch of science. It affects most of our lives on a daily basis, if only in the decisions we make about what to wear before we leave the house. We don't need any specialist equipment to tell us whether it's warm or cold, wet or dry and this sets the science of weather apart from many other branches of science. To be able to observe and fully understand nuclear physics say, or cell biology, we'd need extensive laboratory equipment costing millions of pounds. In order to take our own measurements of the weather, a very modest investment in an electronic weather station package can set us up with all the basic information we need.

Having said that, the atmosphere itself is extremely complex and to fully understand how it works demands knowledge of a wide range of scientific areas on a huge range of scales. At the very large scale it is the position of the Earth in our solar system that sets the temperature at a level that makes the planet habitable. The geometry of the Earth's orbit about the Sun and the tilt of its axis of rotation sets the progression of the seasons. The chemical composition of the atmosphere further determines the temperature of the Earth's surface and creates the conditions that make the Earth a suitable place for life to develop. The laws of motion developed by Isaac Newton and the physics of temperature and heat encapsulated in the laws of thermodynamics describe how and why the atmosphere moves and thus generates the weather systems that affect the planet. On the microscopic scale the interaction of water vapour with minute particles of solid matter determine where clouds will form and whether those clouds will go on to produce rain. On very local scales the interaction of the atmosphere with variations in the Earth's surface geography and vegetation can mean that the weather can vary significantly over very short distances.

Given all this complexity it is really quite amazing that we are actually able to forecast, with a considerable degree of success, how the weather will develop over a period of a few days. Criticizing the weather forecast is something of a universal human trait, but careful analysis shows that today's weather forecasts are really rather good. The fact that we can forecast the weather tells us that we do understand the many branches of science that come together within the atmosphere. In this book we will start off by looking at the very large-scale factors that create the climate of the Earth and the global-scale circulation of winds. Then we shall start to zoom in to look at how individual weather systems form and develop, and how they may be affected by very local conditions such as the presence of hills and mountains, coasts and lakes, cities and changes in natural vegetation. Knowing something about the actual process of producing a weather forecast, from the collection of weather observations through to the final delivery of the forecast, can also help us to understand the forecast better. If we understand what the forecasters are actually trying to tell us, and some of the reasons why the forecast may go wrong, we are able to make intelligent use of the forecast.

Armed with knowledge of the workings of the atmosphere we can even start to produce our own weather forecasts and have a go at beating the professionals. Averaged over any reasonable period of time it is unlikely that we ever will be better at forecasting than a large weather forecasting agency with a staff of hundreds, a budget of millions and a supercomputer able to perform billions of calculations per second. By really understanding the local conditions that make our own weather patterns unique we may be able to add some value to a weather forecast for a large region produced by a major forecasting centre.

Whatever our interest in the weather, whether it be because our work or our hobbies are affected by it, or simply that we find the development of weather patterns and the evolution of the sky above us fascinating, knowing more about how it works can really enhance our appreciation of the complexity of the atmosphere and our ability to make full use of weather forecasts.

1

Introduction

In this chapter you will learn:
- *why the weather is important to life on Earth*
- *what makes the weather a fascinating subject to study*
- *what topics this book will cover.*

The weather is something that influences most of our lives on a daily basis. If we step outdoors it's probably the first thing we notice, even if only semiconsciously. Even before we go outside, the weather has probably influenced our decisions about what to wear and perhaps what we are going to do. In today's modern world where we can pretty well insulate ourselves from the effects of the weather, we still notice the state of the weather and perhaps use it as a conversation topic. On a more serious level, the weather affects many things that we perhaps don't think about on a daily basis. The crops that are grown around the world have been chosen for their suitability to a particular climate, and particularly unusual weather conditions during a growing season may affect their productivity. The movement of people and goods around the world, particularly by air and sea, is sensitive to the weather. In some cases, the weather can even be a killer, with strong winds, flooding, drought, and extreme heat or cold all claiming a significant number of lives every year, right around the globe.

Meteorology is a fascinating topic. It brings together many different branches of science in an attempt to understand how the atmosphere works and, perhaps most importantly, to try to forecast how it will evolve in the future, on timescales of hours to decades. In this sense meteorology is unique because, several times a day, meteorological experts in the major weather-forecasting centres around the world put their knowledge to the test by forecasting the weather and then broadcasting the results. No other scientific discipline puts itself to public scrutiny in quite the same way, and this constant testing is producing results. Forecasts of the weather three days ahead are currently more accurate than forecasts only one day ahead would have been 25 years ago. And although there are still occasions when weather forecasts are wrong, even at quite short ranges, much of the criticism levelled at weather forecasters these days is unjustified. But however good weather forecasts become, it seems to be part of our human nature to criticize them.

Perhaps one of the things that makes the weather interesting is that, at its most basic level, we don't need any special equipment to observe it. Our bodies will let us know whether it is hot or cold, windy or calm, humid or dry, and obviously we can tell whether there is any precipitation or whether the Sun is shining. This allows us to make a subjective judgement of how good the weather forecast was. If we become really interested in the subject, we can buy scientific instruments such as thermometers or barometers pretty cheaply. For a fairly modest sum of money we can even buy a complete automatic weather station that we can set up in our garden, link to a computer and produce continuous records of the weather for our home location. With access to some simple weather instruments we can all become scientists, testing the quality of the weather forecast on a daily basis. The next logical step is then to see if we can do any better ourselves, perhaps by applying our knowledge of local conditions that affect the weather in a way that a forecast provided by a national meteorological agency is unable to do.

Another aspect of the weather that has become big news over the last 20 years or so is climate change. Most people are now aware that the Earth's climate is undergoing changes that are outside the limits of natural variability. Every time that an extreme weather event such as flooding, drought or maybe a particularly intense hurricane, tornado or storm makes the news, the question of whether climate change is somehow 'to blame' is raised. The issue of climate change has added a new dimension to meteorology because now, as well as trying to predict the weather a few days ahead, scientists are being asked to predict how weather patterns will change over the coming decades and even centuries. Some of the changes could affect the lives and livelihoods of many millions of people and so there is a great deal of pressure to improve our knowledge of how the atmosphere will respond to increasing levels of carbon dioxide and other gases.

This book will look at how the atmosphere behaves and why. It will start by looking at the nature of the atmosphere, and the broad pattern of climate and weather patterns on a global scale. This will set the scene for the following chapters which will look in more detail at the nature of the weather on regional and then local scales, examining how individual weather systems work and then considering how local conditions can alter the effects of these weather systems. The conditions that lead to extreme weather events such as hurricanes, tornadoes, floods and droughts will also be considered. The next few chapters will move on to consider how the atmosphere is observed and forecast, both by the big national meteorological agencies and also how individuals might be able to make their own observations and attempt their own forecasts. Finally the subject of climate change will be discussed in the context of some of the most recent research into this important topic.

Whatever your interest in the weather, this book will hopefully serve as a starting point for finding out how the atmosphere works and what makes the weather. Taking it further lists useful books and websites is included so that you can find current weather information, forecasts and educational material to help you take your interest further.

2

A survey of the atmosphere

In this chapter you will learn:
- *how the Sun provides the energy for all our weather*
- *what the Earth's atmosphere is made of*
- *about the different layers of the atmosphere.*

In order for a planet such as the Earth to have any weather as we understand it, two ingredients are necessary. The first of these is an atmosphere – that is a body of gas, held in place above the surface by the planet's gravity. The second ingredient is an energy source that provides the driving force to set the atmosphere in motion. All the weather that we experience is simply a result of motions in the atmosphere, driven ultimately by the energy supplied by the Sun. Before examining the details of specific weather systems, a brief look at these two fundamental ingredients sets the scene upon which each day's weather around the globe unfolds.

The Sun

All the energy that drives atmospheric motion is provided by the Sun. Essentially a flaming ball of hydrogen and helium, the Sun emits energy in the form of electromagnetic radiation which can travel through the vacuum of space. The total energy emitted by any object is related to its surface temperature. Hot objects emit a lot of energy, whereas cold ones emit much less. It is because of this relationship that we are able to estimate the surface temperature of the Sun at about 5,700 degrees Celsius (°C), or 10,300 Fahrenheit (F). The interior of the Sun is very much hotter – probably in excess of 10 million degrees Celsius. The energy emitted by the Sun includes gamma and X-rays, visible light, infrared radiation and even radio wavelengths. The Sun gives out much of its energy in the form of *visible* light. It is no coincidence that our eyes have evolved to see the radiation in these particular wavelengths. If the Sun emitted most of its energy as infrared radiation, our eyes could well have evolved to see these wavelengths instead, giving us a very different view of the world. The Earth also emits energy in the form of radiation, as does any body with a temperature above absolute zero. However, since the Earth is so much cooler than the Sun its wavelength of maximum emission is very much longer – in the infrared part of the spectrum.

The amount of power that the Sun emits is truly massive – about 4.0×10^{26} watts. It is so colossal that it is difficult to think of anything to compare it with. The consumption of electrical power by the whole of the UK on a typical winter evening is about 40 gigawatts, so the Sun is generating about ten thousand million million times this amount of energy. There is nothing in space to absorb this energy. However, as it radiates away from the Sun it is spread over an increasingly large area so the amount of energy *per unit area* reduces with distance. The Earth is about 150 million kilometres away from the Sun but even so it still receives about 1,380 watts per square metre at the top of the atmosphere. This number is often referred to as the 'solar constant'. This is a rather poor description as it isn't constant at all. It varies throughout the year since the Earth's orbit about the Sun is not circular, so the Earth–Sun distance changes throughout the year. The energy emitted by the Sun also varies on a number of other timescales, most notably that of the 11-year sunspot cycle. On vastly longer timescales – tens of thousands of years – changes to the shape of the Earth's orbit around the Sun, variations in the tilt of the Earth's axis of rotation and shifts in the point of the orbit around the Sun at which the equinoxes occur all affect the amount of energy received by the Earth. Of these mechanisms, it seems that the changes to the shape of the orbit have had the largest effect on the Earth's climate. The orbit changes from elliptical to almost circular and back again with a period of about 100,000 years. This cycle seems to correspond well with the occurrence of ice ages and the intervening warm interglacial periods.

The amount of energy per unit area from the Sun arriving at the Earth's surface varies with position on the surface. At the equator the Earth's surface is close to being perpendicular to the incoming rays of the Sun. Therefore the energy in the beam is spread over a relatively small area compared to locations further away from the equator where the angle of incidence of the Sun's rays is much more oblique (see Figure 2.1).

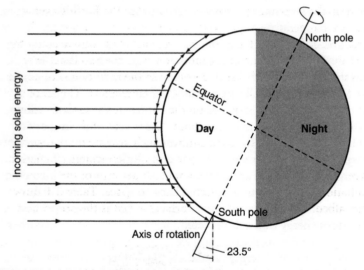

Figure 2.1 Variation with latitude in the amount of energy from the Sun received at the Earth's surface, for the northern hemisphere winter. The black arrows indicate that, as latitude increases, the energy from the Sun is spread over a larger area. Due to the tilt of the Earth's axis, the North Pole is in perpetual darkness at this time of year.

This partly explains why tropical parts of the world are warmer than polar regions. The situation is complicated by the fact that the Earth's axis of rotation is tilted with respect to the plane of the Earth's orbit around the Sun. This tilt, about 23.5° from the perpendicular, is fixed relative to the plane of the orbit so that at some times of the year some parts of the polar regions of the Earth are receiving no energy from the Sun at all. The tilt of the Earth's axis determines the progression of the seasons. In northern hemisphere winter it is the Arctic that receives no energy from the Sun, whereas during northern hemisphere summer the Antarctic is in darkness. If the Earth's axis were exactly perpendicular to the plane of its orbit about the Sun then we wouldn't experience any seasonal variations in our weather.

Broadly speaking the Earth is in a 'radiative equilibrium' with the Sun. That means that the amount of incoming energy received by the Earth from the Sun is balanced by outgoing radiation emitted by the Earth – usually called 'terrestrial' radiation. This allows us to make a very simple calculation of what the Earth's average surface temperature ought to be if there was no atmosphere. In order to make this calculation, we must also take into account the fact that not all the incoming radiation from the Sun is actually absorbed by the Earth's surface. Some of the radiation is reflected back into space without having any warming effect. The amount of radiation reflected depends on the nature of the surface. Snow- or ice-covered surfaces reflect about 90 per cent of the radiation falling on them whereas dark surfaces such as coniferous forests or dark soils only reflect about 20 per cent of the incoming radiation. Averaged over the whole globe, about 30 per cent of the incoming radiation from the Sun is reflected back to space. Table 2.1 shows the 'albedo' of various different surfaces – that is the proportion of the Sun's energy that gets reflected.

Surface	Albedo
Fresh snow	0.95
Old snow	0.5
Desert	0.2–0.5
Forest	0.05–0.2
Grassland	0.15–0.3
Soil	0.05–0.4
Water	0.05–1.0

Table 2.1 Albedo of different surface materials.

There are further complications in calculating the balance of energy between the incoming solar radiation and the outgoing terrestrial radiation. All parts of the Earth's surface are radiating energy out to space all the time. However, at any one time, only half the Earth's surface is receiving any energy from the Sun – the other half is in darkness. Add to that the fact that the Earth's surface is curved and the result is that only an area the size of a flat circle with the Earth's radius is receiving incoming solar radiation. If we take all these things into account then the theoretical temperature of the Earth's surface in the absence of the atmosphere can be worked out. The answer is about –18°C (0 F) which is too cold to support most of the life forms which inhabit the Earth. In actual fact the average surface temperature of the Earth is more like +15°C (59 F), so something must be going on to bring about this difference. In fact, the difference is due to the presence of the atmosphere and the nature of the gases within it.

Composition of the atmosphere

The Earth's atmosphere is composed of a mixture of gases, extending up to about 100 kilometres above the surface. The density of this mixture decreases with height above the surface. At the surface, the density of air is about 1.25 kilograms per cubic metre. *Air pressure*, as measured by a barometer in units of millibars (mb) – see Appendix – is simply the weight of the column of atmosphere above the instrument, so clearly this will also decrease with height. There are many different gases within the atmosphere, but nitrogen at 78 per cent and oxygen at 21 per cent account for most of the total mass. The remaining one per cent or so is a mixture of many different gases, called *trace gases*. This small fraction actually turns out to be the most interesting in terms of influencing the weather, and several of these trace gases punch well above their weight meteorologically speaking. Three in particular have a massive impact on the Earth's climate. These are ozone, carbon dioxide and water vapour. It could be argued that the presence in the atmosphere of these three gases, in the particular concentrations at which they occur, is actually crucial for the existence of human life on Earth.

OZONE

The atmosphere is largely transparent to the Sun's radiation. Clouds have the effect of reflecting solar radiation back out to space, but very little energy from the Sun is actually *absorbed* by the atmosphere. This means that most of the radiation from the Sun gets through to the Earth's surface, at least in clear sky conditions. The major exception to this is that the very short ultraviolet wavelengths are absorbed by atmospheric *ozone*. Ozone is a molecule composed of three atoms of oxygen. This differs from the more common form of atmospheric oxygen that consists of two oxygen atoms. Although present in the atmosphere in tiny amounts – about 12 parts per million – there is enough ozone in the atmosphere to filter out the short wavelengths of solar radiation almost completely. This is a good thing as radiation at these wavelengths can damage human tissue and lead to skin cancer. Most of the ozone in the atmosphere occurs at heights of between 20 and 30 kilometres above the surface, in a region of the atmosphere known as the *ozone layer*. The interaction of solar radiation with ozone molecules, together with a complex series of chemical reactions, means that ozone is constantly being destroyed and recreated. In the natural scheme of things, the amount of ozone within the atmosphere would remain roughly constant. However, in recent years the amount of ozone in the atmosphere, particularly at polar latitudes, has been observed to be decreasing, resulting in the so-called 'ozone hole' over the Antarctic.

Insight – The ozone hole

Every year, at the end of the southern hemisphere winter, ozone amounts in the atmosphere high over the Antarctic are known to reduce. However, in the early 1980s measurements taken by scientists working at Antarctic research stations

(Contd)

showed that the ozone above them appeared to be decreasing much more rapidly than previously observed. At first these measurements were dismissed as faulty, but were found to be correct after careful checking. The reason for the rapid depletion of ozone was found to be chemical reactions with man-made gases called chlorofluorocarbons (CFCs), which were being widely used at the time in the manufacture of refrigeration systems and as propellants in aerosol spray cans. Depletion of ozone is potentially a serious problem since it would allow much more short-wave energy from the Sun to get through to the Earth's surface, increasing skin cancer risk to humans and animals. Rapid international action led to the implementation of the Montreal Protocol in 1989, an agreement banning the use of CFCs and other ozone-destroying chemicals. Since then the depletion of atmospheric ozone seems to have slowed, but it is perhaps too early to say whether ozone levels will ever return to pre-industrial values.

CARBON DIOXIDE

While the atmosphere is largely transparent to solar radiation, the same is not true of the radiation emitted by the Earth's surface. Because the Earth is so much cooler than the Sun, the radiation it emits is at much longer, less energetic wavelengths. While the main atmospheric constituents are pretty well transparent to these wavelengths too, some of the trace gases do absorb some of this radiation. Chief among these so-called greenhouse gases are water vapour and carbon dioxide. Carbon dioxide (CO_2) occurs naturally in the atmosphere in concentrations of between about 200 and 300 parts per million and it is well mixed throughout the atmosphere. It absorbs some of the energy emitted by the Earth, causing a warming of the atmosphere. Some of this absorbed energy is re-radiated back down towards the Earth's surface, keeping the Earth's surface at a higher temperature than it would have if carbon dioxide and the other greenhouse gases were not present in the atmosphere (see Figure 2.2).

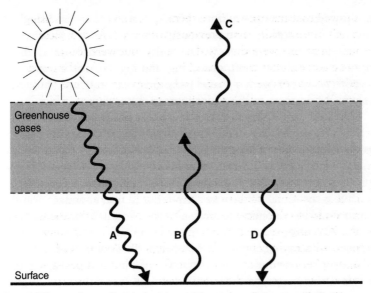

Figure 2.2 The greenhouse effect. In clear sky conditions, solar radiation warms the Earth's surface (A). The Earth radiates energy back upwards (B) and much of this is absorbed by greenhouse gases in the atmosphere. Some energy is then radiated out to space (C) but some is re-radiated back down towards the surface (D), thus warming it further.

In fact this *greenhouse effect* accounts for the 33°C (59 F) warming of the surface described above. It is important to realize that this effect is a naturally occurring phenomenon. It is subject to natural fluctuations, but man-made increases in atmospheric carbon dioxide and other greenhouse gases can lead to an amplification of the greenhouse effect and a warming of the Earth's surface and atmosphere over and above the natural warming. The current level of carbon dioxide in the atmosphere is around 380 parts per million and has been rising steadily over the last 200 years or so. See Chapter 10 for more information on carbon dioxide measurements.

WATER VAPOUR

As mentioned above, water vapour is a greenhouse gas. However, to a meteorologist, perhaps the most interesting property of atmospheric water vapour is that it can change state very easily from vapour to liquid, or even solid, water. The temperature of the atmosphere is such that all three phases of water (vapour, liquid and ice) can occur in close proximity and can easily transform from one phase to another. Water vapour is the source for all the atmosphere's clouds, rain, snow, mist and fog. Forecasting the Earth's weather is such a challenging and fascinating task, largely due to the presence in the atmosphere of water vapour.

The source of virtually all the water vapour within the atmosphere is the Earth's surface, largely the oceans. A tiny amount of water vapour is created within the atmosphere by chemical reactions, but in terms of the weather this is completely insignificant. Because the surface is the source of the water vapour, it is most abundant near the surface and its concentration decreases with height. Near the surface, in the warm tropical regions, there can be up to 30 grams of water vapour in each kilogram of air. At a height of about ten kilometres above the surface this drops to a fraction of a gram in each kilogram of air. There are several different ways of measuring the water vapour content of the atmosphere. The *specific humidity* is simply a measure of the mass of water vapour in each kilogram of air. However, this is not necessarily the most useful measure. Much more informative is the *relative humidity*. This is a measure of how close the air is to being saturated with water vapour. At any given temperature, there can only be so much water vapour present in air before some of the vapour starts to condense into liquid. At this point the air is said to be saturated, and has a relative humidity of 100 per cent. Air that contains no water at all has a relative humidity of zero per cent. At heights of several kilometres above the Earth's surface, air may have a very low *specific* humidity, containing only a few tenths of a gram of water vapour per kilogram, but because the air at these heights is very cold, the *relative humidity* can still reach 100 per cent.

How do clouds and rain form?

In order for clouds and rain to occur, the water vapour that is always present in the atmosphere has to start condensing. This means that the *relative humidity* needs to reach 100 per cent so that the air becomes saturated. There are two main ways for this to happen. The first would be for more water vapour to be added to the air. This happens, for instance, when winds blow unsaturated air across an ocean surface so that evaporation from the ocean increases the water vapour in the air. This would be more likely to form fog. The other way to bring about condensation is to cool the air without changing its water vapour content. Air with relative humidity of 75 per cent at 20°C (68 F) will become saturated if cooled to 15°C (59 F). The temperature at which air would become saturated if cooled is called the *dewpoint temperature*. By far the most common way of air being cooled to its dewpoint is by rising motion of the air – rising air will always cool. Hence weather forecasters are always interested in identifying parts of the atmosphere where the air is rising since these are the likely regions for the formation of clouds and then rain. Throughout the next few chapters we will be looking at some of the different ways that air can be made to rise, thus cooling and forming clouds.

When water vapour begins to condense near the Earth's surface, say on a cool evening, it settles onto objects on the surface forming dew. However, away from the surface there are apparently no solid surfaces for the water to condense onto, so one might expect that the tiny water droplets that make up clouds would be formed of pure water. This is not the case. Water really does need surfaces on which to condense. In the atmosphere, these surfaces are provided by microscopic solid particles, known as *aerosol*. These particles get into the atmosphere from a large variety of sources – wind-blown dust, volcanic eruptions, salt particles from the ocean surface, plant pollens, natural and man-made fires and man-made pollutants. Wherever water vapour begins to condense in the atmosphere, there are always enough aerosol particles to act as sites for the condensation to occur. When condensation occurs, the liquid water molecules cluster around aerosol particles, which are now referred to as *cloud condensation nuclei (CCN)*. If there is a very plentiful supply of these CCN, over a polluted industrial region for instance, then the water molecules will be divided between many nuclei and so each individual cloud droplet will be very small. If the CCN are less plentiful, the same number of water molecules will be divided between fewer nuclei and so each individual cloud droplet will be larger. This can affect the rate at which precipitation develops. In relatively clean air with fewer CCN, clouds will produce rain more readily than in polluted air because the cloud droplets grow large and so fall out of the cloud due to gravity more quickly than the smaller drops in the polluted air.

The vertical structure of the atmosphere

Because the atmosphere is largely transparent to the Sun's radiation, most of that radiation gets through to the surface of the Earth. Some of this is reflected back out to space, but averaged around the globe, about 70 per cent of energy from the Sun is absorbed by the surface and so it warms up. Thus the Earth's surface is the major source of heat for the atmosphere. The temperature generally decreases with height away from the surface, in the lowest ten kilometres or so. However, in some parts of the atmosphere, the temperature is observed to stay constant or even increase with altitude. This situation is called a *temperature inversion*. Changes in the observed temperature vary with height, or *lapse rate*, meaning that the atmosphere can be divided into four distinct vertical layers (see Figure 2.3).

THE TROPOSPHERE AND VERTICAL MOTION

The troposphere is the layer of atmosphere closest to the Earth's surface, extending up to about 15 kilometres in the warm tropical latitudes, and perhaps 8–10 kilometres in higher, cooler latitudes. It has two main characteristics. The first is that the temperature *generally* decreases with height at an average rate of about 6°C (11 F) per kilometre. However, near the surface, particularly on calm, clear nights, the temperature can increase with height as the surface cools rapidly by radiation to space and the layers of air directly above the surface also cool by contact

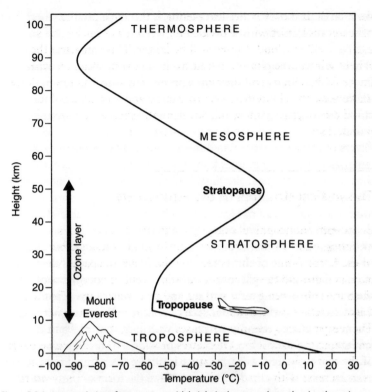

Figure 2.3 Average variation of temperature with height in the atmosphere, showing the main atmospheric layers.

with the cold surface. The second distinguishing feature of the troposphere is that it contains almost all the water vapour in the atmosphere. Thus almost all the weather that we experience at the surface in terms of cloud and precipitation forms within the troposphere. Vertical motions of air are common in the troposphere and can be very large. In fact, the name troposphere was coined from the Greek for 'turning sphere' because of the vertical overturning circulations that mix up the air in this layer. Vertical motions in the atmosphere are of real interest to meteorologists and weather forecasters. As air rises through the atmosphere it cools, and this cooling can lead to the condensation of water vapour, forming clouds and potentially precipitation.

So a large part of weather forecasting is trying to predict where and when vertical motion will occur.

There are two main ways in which significant upward vertical motion of air occurs in the troposphere. The first is gradual, slantwise vertical motion, with air rising at rates of perhaps only a few millimetres per second, occurring over areas of perhaps hundreds or even thousands of kilometres. This leads to widespread layers of clouds and associated precipitation. The second way that air rises is in vertical plumes known as *convection*. This occurs most often when the air near the Earth's surface is heated by contact with the surface itself, perhaps on a sunny day or when cool air is blown over a warmer ocean or land surface. If a parcel or bubble of air is warmed enough it will become buoyant and will rise. Since air pressure reduces with height the parcel will expand and cool as it rises. At each subsequent level, if the parcel is still warmer than the air around it then it will continue to rise, but if it becomes cooler than the surrounding air it will start to sink back down again. If condensation of water vapour starts to occur within the parcel then latent heat will be released which will warm the parcel – partly counteracting the cooling effect of expansion and making it more likely that the parcel will continue to rise. At this point cloud will start to form. This cloud will initially take the form of puffy white 'fair weather' clouds, but depending on the exact rate at which temperature changes with height in the atmosphere and the amount of water vapour in the rising air parcel, the cloud may develop rapidly into towering clouds which may even start to produce showery rain. Further deepening may even lead to thunder and lightning.

THE STRATOSPHERE

In the early twentieth century, several meteorologists developed instruments that could be carried up through the atmosphere on balloons, making a record of the temperature and humidity as they ascended. It was noted that beyond about ten kilometres above the surface, the temperature stopped decreasing with height, and actually began to increase. This layer of the atmosphere, which extends up to about 50 kilometres, became known as the stratosphere, since the temperature structure effectively inhibits vertical motions so that the atmosphere at this level is very *stratified* with little vertical mixing occurring. The reason for the increasing temperature is the presence in this layer of the majority of the atmosphere's ozone. As discussed earlier, this gas absorbs solar radiation and so heats the atmosphere. The boundary between the troposphere and the stratosphere is known as the *tropopause*. The tropopause acts as something of a lid to the troposphere, and exchange of air between the troposphere and stratosphere is limited and occurs in very specific locations. Because of this, very little water vapour gets into the stratosphere and so concentrations of water vapour in the stratosphere are very low indeed.

THE MESOSPHERE AND THERMOSPHERE

At an altitude of about 50 kilometres above the surface, temperatures once again start to decrease with height. This occurs at a point known as the *stratopause*, and above this level is a layer known as the *mesosphere*, literally 'middle sphere'. Typical temperatures at the stratopause are around 0°C (32 F) and decrease to around –100°C (–148 F) at an altitude of about 85 kilometres. At this point the density of the atmosphere is about 1/10,000th of its surface value. Above this level, temperatures begin to increase again as some of the lighter atmospheric gases here absorb solar radiation. Temperatures can increase to above 1,000°C (about 1,800 F), but because the atmosphere here is so thin, the actual definition of temperature becomes a little meaningless. This upper layer of the atmosphere is known as the *thermosphere*.

Summary

The composition of the atmosphere is a crucial factor in allowing the Earth to support life. The presence of various gases within the atmosphere means that the surface of the Earth is warmer than it would be with no atmosphere and that harmful ultraviolet radiation from the Sun is filtered out before it reaches the surface. The constant re-distribution of water around the globe by the atmosphere in the form of precipitation means land regions can support life due to the availability of fresh water. The structure of the atmosphere and the distribution of energy from the Sun over the Earth's surface lead to the large-scale patterns of air motions around the atmosphere. These large-scale motions effectively determine the climate and the particular nature of weather systems experienced at any particular point on the globe. In the next two chapters we'll look at the large-scale wind patterns and smaller-scale weather systems that occur as a result of the uneven distribution of the Sun's energy across the surface of the Earth.

10 THINGS TO REMEMBER

1 *All the weather on the planet is caused by the atmosphere being in motion.*

2 *All atmospheric motion, and hence all our weather, is driven by energy received from the Sun.*

3 *The tilt of the Earth's axis relative to the plane of its orbit around the Sun leads to the seasons.*

4 *Bright surfaces such as snow and ice reflect the Sun's energy back out to space much more effectively than dark surfaces such as soil or the ocean surface.*

5 *Atmospheric pressure is simply the weight of the column of atmosphere above your head pushing down on you due to gravity.*

6 *The Earth's atmosphere is composed mostly of nitrogen (78%) and oxygen (21%).*

7 *It is the presence of minute quantities of other gases which make the Earth suitable for life.*

8 *Ozone filters out harmful ultra-violet radiation from the Sun before it can reach the surface.*

9 *Carbon dioxide and various other 'Greenhouse Gases' in the atmosphere have the effect of warming the atmosphere so that the average temperature of the Earth's surface stays above freezing.*

10 *The presence of water vapour in the atmosphere allows clouds and rain to form and so is a crucial element in the generation of weather.*

3

The global circulation

In this chapter you will learn:
- *why the Earth's atmosphere is in constant motion*
- *how the Earth's rotation affects the winds*
- *about the typical patterns of winds and weather across the globe.*

In the previous chapter we saw how the tropical regions of the Earth receive more energy from the Sun than polar regions. This explains why the tropics – that part of the Earth's surface near to the equator – are warmer on average than the poles. By thinking a bit more about this situation, we'll start to uncover why the atmosphere is in constant motion and why we have weather at all.

The equatorial regions of the Earth receive more energy from the Sun than the polar regions as we have already seen. The Earth's surface is also radiating energy back to space, and averaged over the whole Earth and over a reasonable length of time this outgoing radiation balances the incoming. The equatorial regions of the Earth radiate more energy to space than the polar regions because the surface near the equator is warmer, but the variation of outgoing energy with latitude is much smaller than the variation of incoming energy, as shown in Figure 3.1.

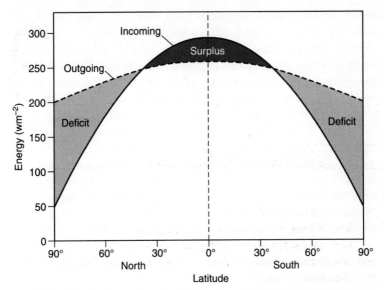

Figure 3.1 Graph showing the energy surplus in the tropics and deficit in higher latitudes due to the difference between incoming solar radiation (solid line) and outgoing terrestrial radiation (dashed line).

A consequence of this difference is that, averaged over time, the equatorial regions of the Earth experience a net surplus of energy, whereas the polar regions experience a net loss. So why aren't the tropical regions constantly getting warmer while the polar regions get colder and colder? The answer is very simple but it has profound consequences for the way in which the atmosphere moves. The atmosphere and the oceans act to even out this imbalance, transporting heat away from the equator and towards the poles. Much of the motion that occurs in the atmosphere, and many of the weather systems that affect the globe, are simply a by-product of this poleward transport of heat by the atmosphere.

The Hadley circulation

On the global scale of things, the motion of the atmosphere is easy to understand. The air near the Earth's surface throughout the tropical regions is very warm, so it rises. As it rises it cools, and as it reaches the level of the tropopause it is prevented from rising further so it flows polewards. At some latitude it sinks and completes the circuit by flowing back towards the equator near the surface. The same sort of circulation of air would occur if you put a heater in the centre of a cold room. The air immediately above the heater would be warmed, and so it rises. On reaching the ceiling the air can't rise any more, so will spread out across the ceiling. As it moves away from the centre of the room it will be cooling down as it is no longer close to the heater, and so will start to sink back down towards the floor. The rising air near the heater needs to be continually replaced, so air from near the edges of the room will flow towards the heater just above the floor. This will carry on as long as the heater is switched on, with warm air rising in the centre of the room and cool air sinking down towards the floor near the walls and then flowing back towards the heater just above the floor. There are a number of factors which mean that the global wind patterns of the atmosphere are actually much more complex than this simple picture, but this simplest possible case is a good starting point for understanding winds on the global scale. The surface part of this circulation, flowing into the tropical regions from both sides of the equator, results in fairly steady winds across the oceans that sailors have known about for a long time. These so-called *trade winds* were relied upon to take the trading vessels of the seventeenth, eighteenth and nineteenth centuries along the global trading routes that were developing at that time. The complete circulation pattern, with air rising over the warm Earth's surface near the equator, flowing towards the poles in the upper troposphere, sinking back to the surface and then flowing back towards the equator as trade winds at the Earth's surface, is known as the *Hadley circulation* or *Hadley cell*, after the British scientist **George Hadley** (see Figure 3.2).

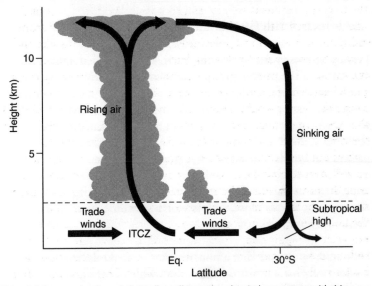

Figure 3.2 Cross section through the Hadley cell in northern hemisphere summer, with rising motion near the equator, sinking motion in the subtropics and the surface trade winds.

Insight – Mapping the trade winds

Edmond Halley (1656–1742), the British astronomer and oceanographer, was the first person to publish a map of the trade winds in 1686. He realized that strong heating by the Sun at the equator was the cause of these winds. However, he tends to get very little credit for this work, being best known for his work on the comet that bears his name. It was another British scientist working half a century later who actually has this pattern of winds named after him. **George Hadley** (1685–1768) wanted to explain why the trade winds blow, not directly towards the equator, but instead have a pronounced easterly component. He proposed that the Earth's rotation would result in a deflection of the winds towards the west, accounting for the observed pattern. We now know that these winds vary considerably in strength at different locations around the equator and with time of year. However, the work of these early meteorologists laid the basis for our understanding of how the atmosphere works on a global scale.

The effect of the Earth's rotation

Hadley proposed that it was the Earth's rotation that caused the deflection of the trade winds giving them an easterly component, and he was correct. In fact the rotation of the Earth has profound consequences for understanding all sorts of atmospheric motions, so it is worth giving some more thought to this concept. The rotation of the Earth about its axis makes understanding the patterns of winds that we observe in the atmosphere rather more complicated than if the Earth didn't rotate. However, understanding this effect is crucial to the understanding of what direction the winds blow, so it is worth spending some time thinking about how it works.

Let's start by considering a non-weather related example. A pilot wishes to fly his aircraft from London, which is on the Greenwich meridian at 51 degrees north to Accra in Ghana, which is also on the Greenwich meridian at about six degrees north. So he takes off and points the nose of his plane due south. After flying for several hours he arrives at six degrees north, only to discover that the Earth has rotated beneath him as he travelled, and he's not over the Greenwich meridian anymore but is instead somewhere over the Atlantic Ocean or South America. Of course, no pilot would be foolish enough to neglect the Earth's rotation when planning a route like this, but the example illustrates an important point. If there were an observer out in space, sitting stationary above the rotating Earth, they would see the plane's track as indeed being a straight line. But to anyone on Earth, including the rather disgruntled passengers in the plane, the plane's track has curved to the right, just as the trade winds in the northern hemisphere are deflected to the right as they blow towards the equator (see Figure 3.3).

Now if we replace the aircraft by a parcel of air moving around in the atmosphere with the wind, the effect of the Earth's rotation will be exactly the same. Although the winds blow at speeds which are generally quite a bit less than the speeds at which a commercial

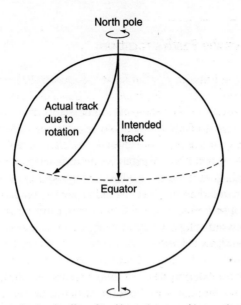

Figure 3.3 Schematic showing the effect of Earth's rotation on an object moving from the North Pole towards the equator.

airliner flies, the winds will still appear to be deflected from a straight line by the effect of the Earth rotating beneath them. The important result is this: in the northern hemisphere, no matter which direction air is moving initially, the winds will always be deflected to the right by the effect of the Earth's rotation. In the southern hemisphere, the winds are always deflected to the left of their original direction by this same effect.

This effect of the Earth's rotation on atmospheric motion is known as the Coriolis effect after the nineteenth-century French engineer **Gaspard-Gustave Coriolis** who wrote down the mathematical framework which describes it. However, Coriolis was concerned with motion in rotating engineering systems such as turbines and waterwheels, and none of his own work applied the principle to the atmosphere. Perhaps it ought to be named the Hadley effect since it was George Hadley who noted that the trade winds would be deflected towards the east as they blow towards the equator.

The importance of the Earth's rotation

The rotation of the Earth means that the trajectory of any object in motion above the Earth's surface will appear to be deflected as the Earth rotates beneath it. However, this effect is really only significant if the motion itself extends over quite a large distance. For instance, a golfer lining up a drive or a footballer kicking a ball to a team mate at the other end of the pitch doesn't need to worry about allowing for the rotation of the Earth. However, an artillery crew aiming a missile at a target several tens of kilometres away would need to start accounting for the Earth's rotation, since over this distance the Coriolis effect will mean that the missile will land some distance away from its intended target. Pilots flying commercial routes over thousands of kilometres certainly have to account for the Earth's rotation as we have already seen. In weather terms, the same is true – the Coriolis effect only becomes important on scales in excess of a few tens of kilometres. The reason that weather systems such as hurricanes rotate is due to the Coriolis effect, but it is not the main reason for the rotation of tornadoes which are much smaller. And contrary to popular myth, the Coriolis effect has nothing whatsoever to do with the direction in which the water spins when the plug is removed from a bath full of water!

The picture so far

We can now start to build up a picture of the global movements of wind across the Earth. So far we know that the atmosphere is transporting heat away from the equator in a circulation which has rising air in the tropical regions, a polewards flow at upper levels, sinking air at some as yet undetermined latitude and a return flow towards the equator near the surface. This circulation is subject to the Coriolis effect due to the rotation of the Earth, so that the winds in the northern hemisphere are deflected to the right of their original trajectories, while winds in the southern hemisphere are deflected to the left.

The latitude at which the Hadley circulation has its descending branch varies with hemisphere, time of year and factors such as the positions of the continents. On average however, the descending branch occurs at about 30 degrees of latitude in both hemispheres. The position of the descending branch of the Hadley Cell coincides with the locations of the Earth's major deserts. This is because air warms as it sinks through the atmosphere, meaning that its relative humidity decreases. Hence deep, rain-bearing clouds are rare at these latitudes.

At the beginning of this section, we said that the atmosphere must be transporting heat from the equator to the poles so that the equator isn't constantly warming and the poles constantly cooling. But if the Hadley circulation only extends from the equator to about 30 degrees north and south as it is observed to do, it isn't directly taking warm air to the poles. So something else must be happening to ensure that the polar regions aren't constantly losing heat. We'll now go on to think about how heat is transported from the subtropical latitudes further towards the poles.

Mid-latitude waves and the jet stream

We know that the tropical regions are hot and the polar regions are cold, so that there must be a *gradient* of temperature between the two. This gradient isn't uniform with latitude. Between the equator and about 35 degrees north and south, the Earth's surface is fairly uniformly warm, particularly if we neglect contrasts between oceans and continents. Polewards of these regions, the surface tends to be much cooler and so the pole-to-equator temperature gradient tends to be largest at latitudes of 40–60 degrees. A consequence of this concentrated temperature gradient is that in the upper troposphere, there tends to be a belt of very strong westerly winds extending around the globe in each hemisphere.

The winds tend to be strongest in the winter hemisphere (i.e. the northern hemisphere in December and the southern hemisphere in June). Why do these strong winds occur?

ATMOSPHERIC PRESSURE AND WINDS

Before we can look into this, there is an important meteorological concept that we need to get to grips with – that of *atmospheric pressure*, often just referred to as 'pressure'. You may already be familiar with the idea that areas of low atmospheric pressure are associated with bad weather, whereas areas of high pressure are generally associated with good weather. We'll look into why this is in detail in Chapter 4, but first it is important to understand exactly what atmospheric pressure is, and why it is important. In Chapter 2 it was stated that atmospheric pressure is simply the weight of the column of air above a particular point in the Earth's atmosphere. At a point on the surface of the Earth, the pressure is the weight of the whole atmosphere above that point, from the surface right up to the top of the atmosphere. If we go up through the atmosphere the pressure must decrease, simply because there is less atmosphere above us exerting downward pressure. At the very top of the atmosphere, the pressure must be zero. So for two points on the Earth's surface, if there is less air in the column above one point than the other, the pressure, as measured by a barometer (see Chapter 7), will be less.

What makes the wind blow?

Up to this point, we've talked about wind blowing in various contexts, but we haven't discussed the fundamental reason why the wind blows in the first place. The answer lies in differences in atmospheric pressure across the surface of the Earth. Air is a fluid, and so can flow just like water. So if there is less air in a particular column of the atmosphere than in a neighbouring one, air will flow from the column with the greater amount of air (i.e. the higher pressure) towards the column with less air (i.e. lower pressure) to try to even out the imbalance. In fact the difference in air pressure acts as a force, pushing air from high pressure towards low pressure.

So in general terms, winds will blow from areas of higher atmospheric pressure towards areas of lower atmospheric pressure. The Coriolis effect that we've just discussed will then act to deflect this flow (towards the right in the northern hemisphere), but in very simple terms, air wants to move from areas of high pressure towards areas of low pressure. This is one of the main reasons why meteorologists are so interested in looking at pressure maps.

Now let's get back to that strong gradient of temperature between the warm subtropical regions and the colder polar regions. Consider two columns of air in the atmosphere, either side of the strong mid-latitude temperature gradient. On the equatorward side, the air is warm and so is less dense. The air in the column on the poleward side is colder and therefore denser. Because the cold air is denser, the pressure decreases more rapidly with height in the cold column than in the warm column. In other words, you don't have to go up so high in the cold column before the pressure has decreased by a given amount. This means that as well as the temperature gradient between the two columns there is also a *pressure* gradient – i.e. at some height above the surface the air pressure in the warm column will be higher than the air pressure in the cold column. We have already seen that air will move from higher pressure towards lower pressure. This means that high above the Earth's surface, the air is moving from the warm column towards the cold column, as shown in Figure 3.4.

But the Coriolis effect will deflect this motion, and this deflection is crucial to explaining the winds that actually occur. In the northern hemisphere, the air that is moving from south to north will be deflected towards the right, which in this case means towards the east. The end result is that air which was originally moving from south to north actually ends up moving from west to east. Thus we see a rather narrow band of fast moving, *westerly* (i.e. from the west) winds at a height of maybe 7–10 kilometres above the surface. This band or belt of strong winds is known to meteorologists as the *jet stream* or often just the *jet*. This is perhaps a misleading name

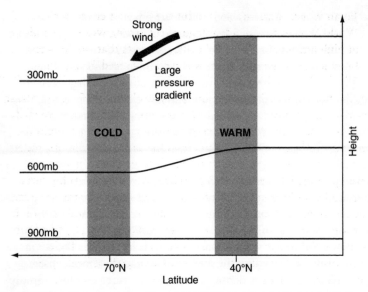

Figure 3.4 Schematic representation of how a horizontal temperature gradient can lead to the formation of a jet stream. The temperature gradient causes a pressure gradient that increases with height. The pressure gradient causes air to move from high to low pressure, and the Earth's rotation then deflects this motion to the right.

as it suggests there is only one, which is perhaps continuous around the globe, whereas in fact there can be several and they are often broken into several segments. Wind speeds in the jet can easily be in excess of 200 kilometres per hour (125 miles per hour).

Insight – Discovery of the jet stream

The discovery of the jet stream(s) is an interesting and relatively recent one. It is often attributed to long-range bombers encountering unexpectedly strong head or tail winds while flying at high altitudes during World War II. However, the term 'jet stream' was coined by the German meteorologist **Heinrich Seilkopf** in the 1930s after taking measurements from instrumented balloons over Europe. But it was a Japanese meteorologist, **Wasaburo Ooishi**, who took the first measurements of strong winds at high altitudes over Japan in the 1920s, using instruments attached to balloons. The fact that Japan launched balloon-borne bombs from

Japan which were intended to hit the US west coast during World War II, taking advantage of the strong westerly winds at high altitudes across the Pacific, suggests that the Japanese knowledge of the jet stream was well advanced at that time.

So, the presence of the jet stream is due to the temperature contrast between air at subtropical latitudes and air a little further towards the poles. But how does the presence of the jet stream explain the *transport* of heat polewards? In much the same way that meanders or eddies can form in a stream of fast-flowing water, *waves* can develop in the jet (see Figure 3.5). These waves can develop large amplitudes and sometimes 'break' in a way shown by the diagram. The jet stream can cut off the neck of a meander, stranding a blob of warm, subtropical air on the poleward side of the jet, or a blob of cold, polar air on the tropical side. This is a pretty efficient mechanism for achieving the poleward transport of heat by the atmosphere.

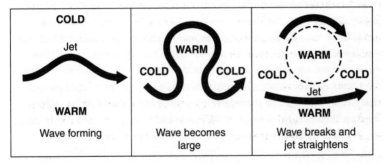

Figure 3.5 A wave in the jet which 'breaks', leading to warm air being stranded at high latitudes.

This is a very simplistic description of what is a very complex process, and a large branch of theoretical meteorology is devoted to understanding the behaviour of these waves and their overall role in the global circulation. However, for practical purposes it is important to know that these waves exist, as they are associated with the 'depressions' and 'fronts' that dominate the weather of much of the mid-latitude regions of the globe and form the familiar features on weather charts that we see in weather forecasts on a daily basis. Chapter 4 will look in more detail at these waves, and how they relate to particular weather systems.

The transport of heat from low to high latitudes is completed by a weak vertical circulation near the poles. Rising warm air in mid-latitudes is transported northwards aloft where it cools and sinks over the poles. A return flow of cold air from the poles towards the mid-latitudes near the surface completes this polar cell. The heat transported in this part of the global circulation is small compared with the transports going on within the Hadley cells and the waves in the jet streams.

Heat transport by the oceans

The world's oceans also play an important role in the transport of heat from equator to poles, which in turn has an impact on the weather around the globe. About half of the total equator-to-pole heat transport is performed by the oceans, both in surface currents driven by the winds and in a pattern of deeper currents known as the *Thermohaline Circulation* (THC). Broadly speaking, warm waters flow towards higher latitudes where they cool. Evaporation of water at the ocean surface by winds at these latitudes not only cools the ocean surface but also makes it more salty, and both effects act to make the water denser. Hence it sinks and the cold water masses formed by this process then return towards the equator as currents deep below the surface of the ocean. The thermohaline circulation has an interesting pattern, with strong differences between the different ocean basins. In the North Atlantic, the poleward transport of warm, near-surface waters is very strong. In the North Pacific there is no such strong transport occurring. This means that the surface waters of the North Atlantic Ocean are warmer than those at the same latitudes in the North Pacific.

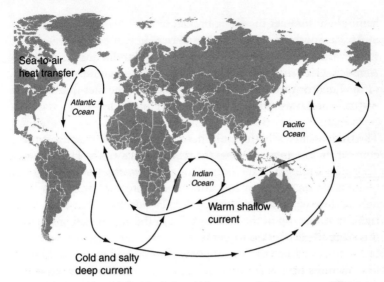

Sea-to-air
heat transfer

Atlantic
Ocean

Pacific
Ocean

Indian
Ocean

Warm shallow
current

Cold and salty
deep current

Figure 3.6 The ocean thermohaline circulation, which transports significant amounts of heat around the globe.

The real global circulation

So far we've discussed the global circulation as if the surface of the Earth was the same everywhere and have neglected the presence of the continents and the progression of the seasons. Of course these things cannot be ignored and they have a profound effect on the climate of the globe. Land surfaces heat up and cool down more quickly in response to seasonal changes in incoming energy from the Sun than do ocean surfaces. Hence the continents tend to be hotter than the oceans during summer (especially in the tropics and subtropics) and cooler than the oceans during the winter (especially at high latitudes). Thus the Hadley cell is not characterized by rising motion of the same strength everywhere around the equator, but by stronger ascent over the warmest surfaces and weaker ascent over cooler areas. The march of the seasons also affects the position of the rising branch of the Hadley cell. In northern hemisphere summer the rising branch moves north of the equator in Africa, South Asia and South America. In winter it migrates southwards, over South Africa, northern Australia and South America. Over the Pacific Ocean, the rising branch remains north of the Equator throughout the year, reflecting the fact that the warmest ocean waters stay north of the equator too.

The trade winds in both hemispheres converge, not exactly at the equator, but in an area below the rising branch of the Hadley cell. This area is known as the *Inter-Tropical Convergence Zone* (ITCZ) and it is characterized by low atmospheric pressure, relatively light surface winds and regions of deep clouds that often extend right up to the top of the troposphere. In the age of sailing ships this region was referred to as the *doldrums*, and sailors dreaded becoming becalmed there as supplies of food and water could run low. Pictures from satellites reveal that the clouds associated with the ITCZ are not continuous around the whole globe, but more often form discrete clumps with gaps in between. Plate 1 shows the Pacific Ocean on a day when the ITCZ did indeed appear as an almost continuous belt of cloud just north of the equator, but closer inspection still reveals a lot of small-scale structure within this band. Further north in the Pacific in this picture are ribbons of cloud associated with meanders in the mid-latitude jet stream.

The descending parts of the Hadley cell, near 30 degrees north and south, tend to be characterized by high surface pressure, clear skies and relatively light winds. Equatorward of these belts of high pressure, the trade winds blow not directly toward the equator but with a marked easterly component due to the Coriolis effect. Polewards, the surface winds have a marked westerly component, again as a consequence of the Coriolis effect. At latitudes of between 40 and 60 degrees north and south, the upper tropospheric jet streams dominate the global circulation. They are associated with bands of low surface pressure, and increased cloudiness. The jet in the southern hemisphere tends to be more continuous around the globe, whereas in the northern hemisphere the presence of the large land masses of North America and Asia, with their associated mountain ranges, means that the jet tends to break up into shorter segments, with the strongest winds appearing over the two major oceans and weaker winds over the continents. Along these jet streams form the weather systems responsible for much of the unsettled weather at these latitudes.

Monsoons

One consequence of the seasonal progression of the maximum in solar heating north and south of the equator, combined with the contrast between ocean and land surfaces, is the occurrence in several parts of the tropics of *monsoons*. 'Monsoon' is derived from the Arabic word for *season*, and it denotes a climate that has marked seasonal variations in wind direction and rainfall amounts. Northern Australia, West Africa and East Asia all experience monsoon climates, but the really notable monsoon is that which affects South Asia and most markedly India. So dramatic is it that it is often referred to by meteorologists as *the* monsoon, with no need to qualify it with the name of the area affected.

During northern hemisphere spring in the vicinity of the Indian Ocean, the warmest surface temperatures and hence the band of clouds and rain associated with the inter-tropical convergence zone (ITCZ), lie over the equator. The Indian subcontinent is heating up rapidly as the Sun moves north of the equator, and the elevated Tibetan Plateau, to the north of the Himalayas, also starts to warm. This rapid warming of the land surfaces has the effect of exaggerating the northwards movement of the ITCZ such that, at some point in late May or early June, the deep tropical clouds move rapidly from the equator to the Indian peninsula. This movement of the clouds and rain is accompanied by change in wind direction and strength, with strong south-westerly winds kicking in across the region. Outbreaks of heavy rain begin over the south of the Indian peninsula, and as June progresses, the rain moves northwards across the country. By the end of June, virtually the whole of India has experienced some downpours. The winds and rain persist through July and August until, during September, the monsoon 'withdraws' from most of the country. Many parts of India receive well over 80 per cent of their annual rainfall during this period and some places on the west coast of India get over three metres (ten feet) of rain in the four-month spell.

During the autumn, the winds over India reverse in direction, becoming north-easterly. The heaviest rainfall ceases, although south-east India and the Himalayan foothills can still get some fairly significant rainfall right through the autumn and winter. Farmers in India rely on the rains of the summer monsoon to water their crops through the growing season, and any crops grown during the winter tend to rely on irrigation from reservoirs filled during the rainy season.

El Niño and the Southern Oscillation

When considering the Earth's weather and climate, it is not possible to think about the atmosphere alone, as the oceans also play an important role. One example of how the oceans and the atmosphere interact to affect the climate is found in the tropical Pacific. It is here that the phenomenon known as *El Niño* occurs.

The changes in ocean temperatures in the East Pacific that are termed El Niño have been known about for several hundred years. Peruvian fishermen noted that, every few years, the fish stocks that thrive in the cold waters along the western coast of South America become severely depleted. This is associated with a considerable warming of the ocean in the region. This occurrence happened regularly enough for the fishermen to give it a name – El Niño, meaning 'boy-child'; a reference to the fact that the warming of the ocean seemed to coincide with Christmas in the years when it happened. The fishermen also noted that often El Niño was followed the next winter by a strong *cooling* of the tropical East Pacific, which they termed La Niña, or 'girl-child'. The Peruvians also noted that, during an El Niño year, the weather they experienced along the coast changed dramatically. Much of this coastal region normally experiences desert-like conditions, but the ocean warming during El Niño was accompanied by heavy rains, often leading to landslides. We now know that these local changes along the Pacific coast of South America are part of a much wider variation in the state of the ocean and atmosphere across the whole tropical Pacific, and that these changes can affect weather patterns on a near-global scale.

To fully understand El Niño and the changes in weather that accompany it, it is important to realize just how closely coupled the ocean and the atmosphere in the tropical Pacific are. Variations in one element profoundly affect the other to the extent that the two cannot be thought of in isolation. It was the great Norwegian meteorologist **Jakob Bjerknes** who, in the 1960s, realized that the key to understanding El Niño lay in understanding the feedbacks between changes in the ocean and changes in the atmosphere over the tropical Pacific. But even today we do not have a complete picture of how El Niño works.

Under normal conditions, the warmest surface waters in the Pacific Ocean are on the western side, between the islands of the Philippines, Indonesia and Papua New Guinea and the Date Line. This warm water also extends downwards to a depth of about 200 metres (660 feet), at which level there is a sharp temperature decrease to much colder water at depth. This sharp gradient is known as the *thermocline*. Moving across the Pacific, from west to east along the equator, the depth of the thermocline is observed to decrease, until, off the coast of South America, it is perhaps only 50 metres (165 feet) below the surface. The water at the surface on this side of the ocean is also much colder than in the west, by about 10°C (18 F). The sea level also varies along the equator. In the west, the mean sea level is about 0.5 metres (1.6 feet) higher than it is in the east.

So why does the tropical Pacific Ocean exhibit this strong east-west gradient in temperature, thermocline depth and sea level?

The answer lies in the atmosphere. Strong easterly trade winds blow across the tropical Pacific for most of the year. These winds push water continually in a current from the eastern to the western side of the ocean. This explains the gradient in the sea level across the basin – if the trade winds stopped blowing, the water which has been effectively 'piled up' on the western side would slosh back towards the east as a wave, and after a period of sloshing backwards and forwards, the surface of the ocean would come to rest in a state with no slope. The winds also explain the gradient in temperature across the basin. Because they are pushing water from east to west, the water in the east has to be replaced by water from elsewhere. This is achieved by *upwelling* of water from deeper in the ocean on the eastern side. This upwelling water is colder than the water at the surface and so the temperature gradient across the Pacific is explained. There is another feedback between atmosphere and ocean that comes into play here. Because the ocean surface is rather cool on the eastern side, the layer of air immediately above it is also cooled. Since the air itself contains a lot of water vapour due to evaporation from the ocean surface, this cooling can cause cloud formation. In fact, much of the tropical East Pacific is covered in a blanket of low cloud for much of the year. This cloud reduces the amount of energy from the Sun that reaches the ocean surface, adding to the cooling effect of the cold upwelling (see Figure 3.7).

So, given that the overlying atmosphere largely sets the conditions in the tropical Pacific Ocean, what sets the atmospheric state? Well, easterly trade winds are common right around the equator, and are caused by the mechanisms described above, i.e. the flow of winds towards the inter-tropical convergence zone, deflected towards the west by the Earth's rotation. However, in the Pacific, this effect is amplified by the temperature gradient across the basin. Because the ocean is warmest in the west, the air over the warm water also warms, and rises rapidly forming deep clouds. This rising warm air must be replaced by air from somewhere else, and in this case the air is drawn in from the central and East Pacific, meaning that the easterly trade winds blow even more intensively. Over on the eastern side of the ocean, the air is generally *descending*. The rising motions in the west and sinking motions in the east

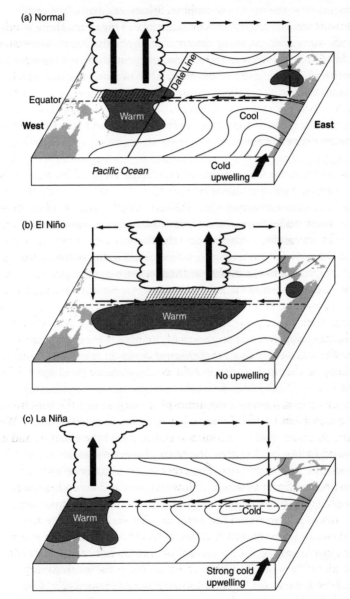

Figure 3.7 Upper level and surface winds, vertical motion, sea surface temperatures and ocean upwelling in the tropical Pacific during (a) normal, (b) El Niño and (c) La Niña conditions.

are linked in the upper troposphere by strong westerly winds, perhaps ten kilometres above the surface. This circulation of air in the vertical plane along the equator above the Pacific Ocean is known as the *Walker circulation*, after **Sir Gilbert Walker**, the British meteorologist and director of the Indian Meteorological Department in the early twentieth century.

So there is a positive feedback between ocean and atmosphere which works like this:

▶ *Easterly trade winds across the Pacific move ocean water from the east towards the west.*
▶ *The water removed from the east is replaced by cold water from below the surface. This results in a strong surface temperature gradient across the Pacific.*
▶ *The strong temperature gradient, with warm water in the west, causes the trade winds to increase in strength.*
▶ *The strong trade winds maintain the temperature gradient in the ocean.*

So what happens to cause the rather sudden warming of the East Pacific ocean every few years that the Peruvian fishermen have known about for so long? Something must break the chain of positive feedbacks. It appears that the key lies in a weakening, or even reversal of the easterly trade winds across the Pacific. The exact cause of this is still an area of active research for meteorologists and ocean scientists, but there are indications that a sustained burst, or series of pulses, of *westerly winds* across the Pacific can sometimes occur that is out of balance with the usual state of affairs where the atmosphere and ocean are tightly coupled. This then causes a response in the ocean. Warm water in the west starts to flow back eastwards because the sea level slope is now not supported by the winds. As this surge of water reaches the eastern side of the basin (and it takes several months to do so), it cuts off the upwelling of cold water along the South American coast and the result is the sudden warming known as El Niño. This warming is not confined to the Peruvian coast but actually extends in a tongue along the equator, as far west as the Date Line. At the same time the ocean in the west cools slightly. This means that the warmest waters in the Pacific are now around the Date Line, and so the rising currents of air which cause deep clouds and heavy rain also move out to be positioned over these warm waters. This movement has a big impact on the atmospheric pressure around the tropical Pacific. Close to the Date Line, pressure falls dramatically, while over northern Australia, there is a large rise in pressure. At the end of an El Niño event, this see-saw in pressure swings back the other way. Gilbert Walker had noted this pressure oscillation while examining weather records from the early twentieth century. He called it the *Southern Oscillation*, but at the time he had no idea that it was intimately linked to changes in the Pacific Ocean.

This movement of the main area of clouds and rain into the central Pacific is the start of the atmospheric response to the changes in the ocean. However, it has much wider consequences. Indonesia, Papua New Guinea and northern Australia become much drier than normal, and severe drought can even result (see Chapter 6 for more details on Australian drought related to El Niño). With the droughts, wildfire risk increases. Rainfall amounts increase on the normally arid west coast of South America, and much of Central America also gets wetter. The Indian monsoon rains are generally weaker than normal during an El Niño event, and the Amazon region of South America also becomes drier. There are also impacts outside the tropical regions. The US north-west and western Canada are usually warmer than normal, whereas California and the US south-east are usually cooler. California also becomes notably wetter than normal. El Niño also has an impact on the frequency and intensity of hurricanes in the Atlantic – see Chapter 6 for more details. The changes in weather patterns associated with El Niño can cost huge amounts of money to the economies of the affected regions. Flooding, drought, wildfires and hurricanes all add up to global costs estimated to be in hundreds of billions of US dollars.

One thing that is sure about any El Niño event is that, at some point, it will come to an end, and the ocean and atmosphere in the Pacific will swing back to their normal states, although often overshooting into the conditions known as La Niña. This is really an *enhanced* normal state, with even colder water than normal waters in the East Pacific and the deep clouds and rain being confined to the extreme west of the basin. So what causes the return to normal conditions? Here, the answer probably lies within the ocean. One simple theory suggests that it's a further consequence of the wave in the ocean that initially travels from west to east to initiate El Niño in the first place. This wave, on hitting the coast of South America, reflects off the coast and travels back towards the west again. As it travels back westward, taking several months to cross the basin, it resets the ocean state back towards its normal state. Cooler water from off the equator may also be encouraged to move to the equator, reversing the warming further. The atmosphere may also play a part in the return to normal conditions. Bursts of strong *easterly* winds may also trigger waves in the ocean that work in the opposite way to the waves triggered by westerly winds that set El Niño going in the first place.

Insight – Forecasting El Niño

The occurrence of El Niño events, and their accompanying shifts in weather patterns, is irregular, happening perhaps every three to seven years. The graph in Figure 3.8 shows the variations of temperature in the Central Pacific over the past 50 years or so, with the strong warm periods, such as those in 1982 and 1997, corresponding to El Niño events. This irregularity makes El Niño difficult to predict. Following the major event in 1982–3, a network of moored buoys was installed across the equatorial Pacific. These gather information on ocean temperatures at and below the surface, ocean currents and surface winds, and transmit the data via satellite links to the major weather forecasting centres. With this information, it is possible to see the early signs of El Niño developing *beneath* the ocean surface. The period 1997–8 saw another major El Niño event, and although this event was bigger than predicted and set in more rapidly than forecast, it was predicted many months in advance.

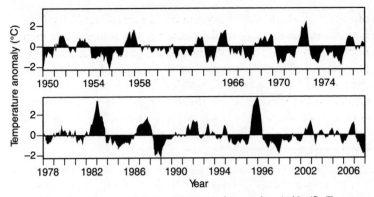

Figure 3.8 Time series of ocean temperature variations in the central tropical Pacific. The strong warmings that occur between every three to seven years correspond to El Niño events.

The global climate

The average weather across the globe is largely a consequence of the global circulation patterns discussed above. The typical weather at any point on the Earth is mainly due to which bit of the global circulation that point happens to be sitting under. Local effects can also have an impact on the weather, and these will be discussed in Chapter 5.

Figures 3.9(a) and 3.9(b) show the average surface air temperatures for the June–August and December–February seasons respectively. These figures show a number of things very clearly. Firstly, the highest temperatures are always over the tropical land masses in the summer hemisphere. In June–August this means that the deserts of North Africa and the Arabian peninsula experience the hottest weather, while in December–February Southern Africa and Northern Australia are the hottest places on Earth. The biggest annual temperature *range*, i.e. the difference between summer and winter temperatures, is experienced by the continental regions of North America and Northern Asia, at latitudes of about 70 degrees north. Here the temperature can change by as much as 50°C (90 F) during the year, with sub-zero winters and warm summers. The smallest annual temperature range tends to be in the tropics,

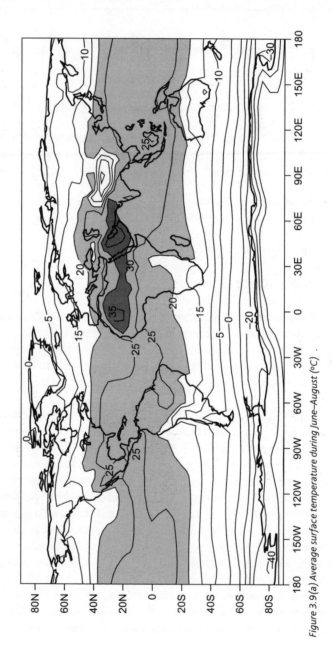

Figure 3.9(a) Average surface temperature during June–August (°C) .

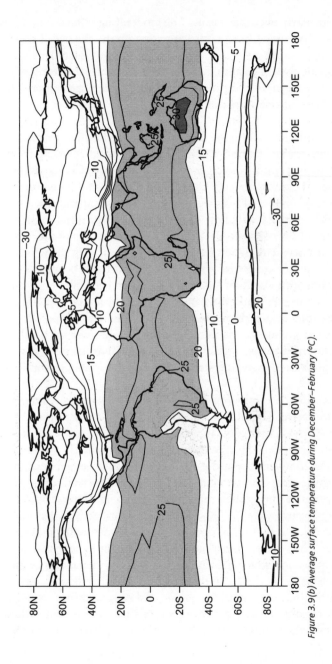

Figure 3.9(b) Average surface temperature during December–February (°C).

particularly over the oceans. The moderating influence of the warm North Atlantic Ocean is also apparent on the climate of western Europe during winter. At the same latitude on the other side of the Atlantic, North-eastern Canada experiences temperatures which on average are well below freezing throughout the winter, whereas western Europe stays much warmer.

The variation in near surface winds throughout the year is shown in Figures 3.10(a) and 3.10(b). The trade winds, flowing towards the inter-tropical convergence zone and westwards across the tropical oceans, are very clear. Note that, during June–August this pattern is replaced over the Indian Ocean by the monsoon flow, in which south-easterly flow south of the equator turns south-westerly north of the equator. Over the subtropical oceans of the summer hemisphere, circulation patterns can be seen, with the winds blowing in a clockwise direction in the northern hemisphere and in an anticlockwise direction in the southern hemisphere. At around 40–60 degrees south the winds are pretty well due westerly and are rather strong throughout the year. This is the region known by ocean-going sailors as the *roaring forties*. Over the northern hemisphere oceans in mid-latitudes, the winds are generally south-westerly, and are stronger during the winter season.

The global precipitation distributions show some very interesting patterns (see Figures 3.11(a) and 3.11(b)). The first thing that stands out is that very large rainfall totals only really occur in the tropical regions. However much someone in the UK for instance might complain about the amount of rain they experience, it is nothing compared to what parts of South Asia or Central America experience during their monsoon seasons. The inter-tropical convergence zone (ITCZ) can be seen very clearly across the Pacific and Atlantic Oceans in both seasons as a narrow band of heavy rain. The very driest regions on Earth are the subtropical latitudes located beneath the descending branches of the Hadley cell in each hemisphere. It is in these regions that the major deserts of the Earth also occur. Both Greenland and Antarctica are also rather dry regions. These places are covered in ice sheets many kilometres thick, which suggests that they ought to experience a lot of snow.

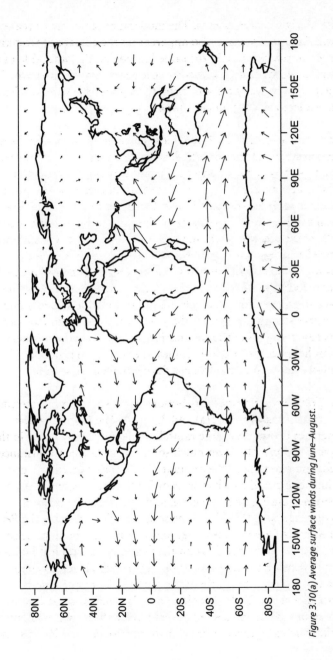

Figure 3.10(a) Average surface winds during June–August.

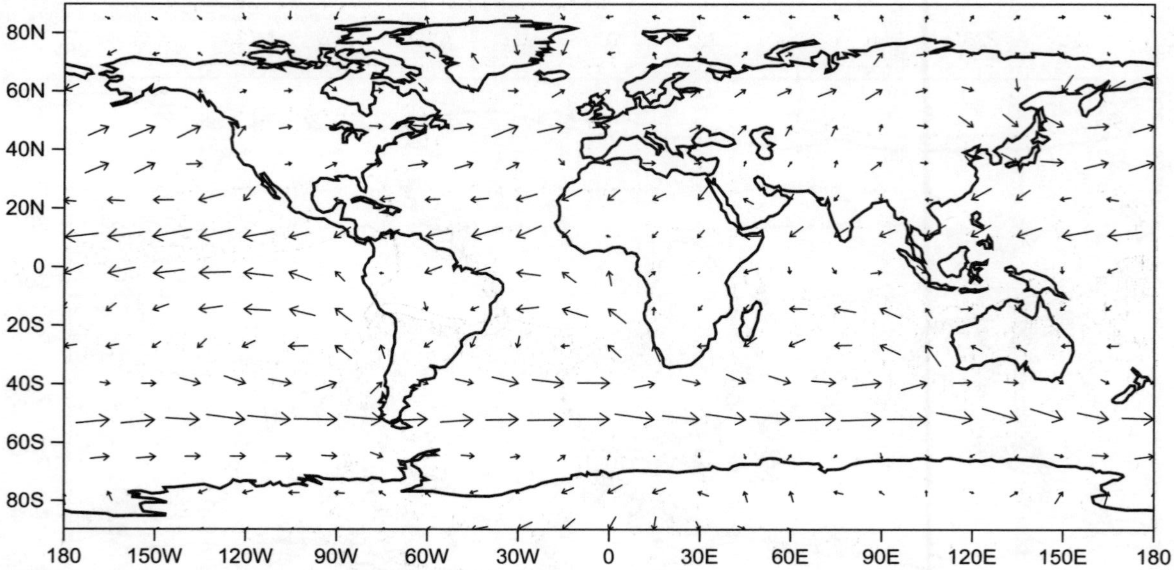

Figure 3.10(b) Average surface winds during December–February.

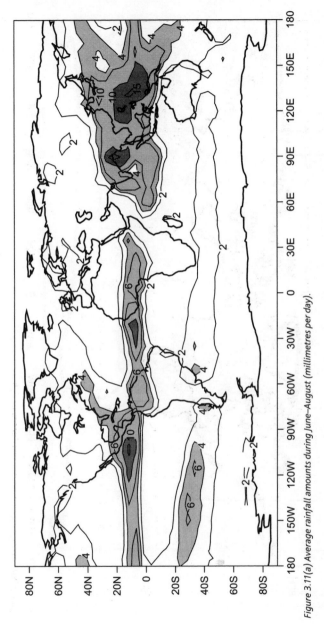

Figure 3.11(a) Average rainfall amounts during June–August (millimetres per day).

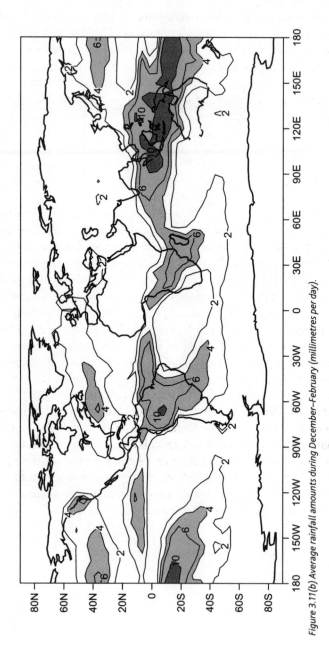

Figure 3.11(b) Average rainfall amounts during December–February (millimetres per day).

However, these ice sheets have accumulated over tens of thousands of years, with only tiny thicknesses of ice being added in any one year by snowfall. The bands of heavier rain across the North Pacific and North Atlantic during the winter months correspond to the position of the *storm tracks* – the typical paths taken by the depressions and associated frontal systems that dominate the weather at these latitudes.

From climate to weather

All of these pictures show the typical picture of the Earth's climate averaged over a number of years. It is important to realize that, on a day-to-day basis, the actual *weather* around the globe can differ considerably from this long-term average picture. The amount of day-to-day variability in the weather is itself quite variable across the globe. In the hot dry desert regions of the subtropics, the weather hardly varies at all from day to day, whereas in the mid-latitude regions between about 40 and 60 degrees of latitude, there can be quite large day-to-day variations in weather with cool, wet days interspersed with dry, warm days. This means that in some regions the weather is more *predictable* – i.e. easier to forecast – than in others. The degree of variability, or to put it in forecasting terms *predictability*, of the weather in any particular region is due to the nature of the specific weather systems that affect that region, and the relative dominance of those particular systems throughout a particular season. In the next chapter we will start to look at the different weather systems that affect different parts of the globe, and think about their consequences on the day-to-day weather conditions in the regions where they occur.

10 THINGS TO REMEMBER

1 *The tropical regions of the globe receive more energy from the Sun than they radiate back to space, whereas the rest of the Earth receives less energy from the Sun than it loses to space.*

2 *The reason that the atmosphere is in motion is that it is trying to even out this imbalance by transporting heat from the tropical regions towards the poles.*

3 *In the tropics, the transport of heat is achieved by the* Hadley circulation, *with warm air rising, flowing towards the poles at high altitudes, sinking at the latitudes of the Earth's major deserts and then flowing back towards the equator near the surface as* trade winds.

4 *The rotation of the Earth deflects the winds towards the right in the Northern Hemisphere and the left in the Southern Hemisphere. This is the* Coriolis Effect.

5 *In the* mid-latitudes *(between about 40 and 70 degrees of latitude) the polewards transport of heat is achieved by waves in the jet stream.*

6 *The world's oceans also transport heat from equator towards the poles in a pattern of currents called the* thermohaline circulation.

7 *Monsoon climates – regions of the world with very marked contrasts between dry and wet seasons – are caused by the seasonal progression of solar heating being enhanced by the presence of large land masses such as South Asia, West Africa and Australia situated polewards of warm tropical oceans.*

8 *El Niño – a warming of the equatorial central and eastern Pacific Ocean which occurs every 3 to 7 years – has profound impacts on weather patterns right around the globe.*

9 *The* climate *of any region of the Earth is a long-term average of the daily weather conditions in that region.*

10 *In some regions the weather is very similar day after day and so is very easy to forecast. In other regions large day-to-day variability of the weather makes forecasting much more challenging.*

4

Weather systems

In this chapter you will learn:
- *what weather systems and air masses are*
- *which weather systems and air masses affect different parts of the globe*
- *why weather forecasters are so interested in pressure maps.*

Chapter 3 described the global-scale circulation patterns that come about as a result of the atmosphere trying to even out the heat imbalance between the equator and the poles. In producing the seasonal average pictures of temperature, winds and rainfall that were shown at the end of Chapter 3, all of the day-to-day variations that we actually think of as 'weather' have been averaged out. Strong winds, thick clouds and persistent heavy rain all tend to occur within quite well-defined and easily identifiable *weather systems*. The nature of these systems varies with location on the globe; the tropics, mid-latitudes and polar regions of the Earth tend to have their own distinct weather systems. These systems are what weather forecasters are observing, tracking and trying to predict on a day-to-day basis.

In this chapter we will look at the various weather systems that affect the different parts of the globe, giving each region its characteristic weather. In effect, these weather systems are the building blocks of the *climate* of a particular region, with the long-term average weather actually being determined by what type of weather systems affect a particular area, and the relative dominance of these different systems through the seasons. Let's think about the average climate of London in January for instance. The average daily maximum temperature is about 7°C (45 F) and the average rainfall in the month is about 75 millimetres (three inches). This doesn't mean that the temperature is 7°C, with 2.4 millimetres of rain every single day in January. Many days may be completely dry, and then there might be a few wet days together. Some days may be much warmer than 7°C, and then it may be quite a bit colder for several days in a row. Occasionally (perhaps once or twice per year) it might be cold enough for snow to fall. These variations will be determined by the particular weather systems that are affecting London on any particular day and so a weather forecaster is far more interested in those weather systems than in the average conditions for London.

Mid-latitude weather systems

Anyone who has spent any time watching televised weather forecasts for the regions of the globe outside of the tropical band will have noticed that forecasters seem to talk a lot about the atmospheric pressure and its effect on the weather. There seems to be a link between high pressure and good weather, and low pressure and bad weather, but we may not understand why this is. In Chapter 3 we saw that differences in atmospheric pressure can act as a force which causes the wind to blow from areas of high to low pressure, and this motion is in turn deflected by the Coriolis effect so that the winds tend to blow not directly from high to low pressure, but parallel to lines of constant pressure. However, this doesn't explain why high pressure often brings good weather, whereas low pressure brings clouds and rain. Very often in the media, weather forecasters will refer to a 'battle' between high pressure and low pressure, with an area of high pressure 'blocking' the movement of a low pressure region, or *depression*, or perhaps an approaching depression 'breaking down' an area of high pressure. This type of language appeals to our imagination but doesn't really help us understand what's actually going on. In this section we will examine why the atmospheric pressure has such a big impact on the weather and look at the different weather systems which are usually characterized by their atmospheric pressure. In order to understand this, we first need to think about how atmospheric pressure is linked to *vertical* air motions and hence clouds and rain.

Atmospheric pressure, vertical motion and the weather

We have already seen in Chapter 2 that atmospheric pressure is simply a measure of the weight of air in the column above a particular point on the Earth's surface. High pressure at the surface of the Earth is caused by there being more air in the vertical column from the surface up to the top of the atmosphere than in a column with lower pressure at the surface. We have also already seen in Chapter 3 that this leads to a flow of air near the surface from high pressure towards low pressure, which is then caused to turn to the right (in the northern hemisphere) by the Coriolis effect. However, we have not yet explained why there was more air in the column with high surface pressure in the first place. The answer often lies in the upper levels of the troposphere, at heights of between five and ten kilometres above the surface, where the winds are often much stronger than they are near the surface. At this level, the winds often blow in such a way that air is arriving at a particular location faster than it is being taken away. This is called 'mass convergence' – the mass of the atmosphere (i.e. the air) is literally *converging* or *gathering into* a particular region. The opposite of this is 'divergence' where the winds are taking air *away* from a particular location faster than it is arriving.

These processes of convergence and divergence explain the changes in atmospheric pressure we measure at the surface. If air is *converging* into the top of an atmospheric column, the pressure at the surface will *rise*. Conversely if air is *diverging* out of the top of the column, the surface pressure will *fall*. At the surface, the result of these changes in pressure will be surface winds taking air from high pressure towards low pressure, as we have already seen, in order to even out the imbalance in surface pressure. But there will also be a crucial third component of air motion involved which links what happens near the top of the column with what happens near the bottom. In a column of air with high atmospheric pressure at the surface, air will be flowing into the column at upper levels and out of the column near the bottom. So the air in the rest of the column must start moving downwards to make up for the air leaving the column at the bottom. If this downwards motion, or *descent*, did not occur, then eventually there would be a vacuum near the bottom of the column because the air flowing out was not being replaced by air from somewhere else. Clearly a vacuum is not possible in the atmosphere, which is a continuous fluid, and there is no other way to replace the air which is leaving the column than by the air above it moving downwards. Figure 4.1(a) shows a simplified representation of this situation.

In the case of low surface pressure, air is *diverging* from the column near the top, and converging into the column near the surface. This means that there must be upwards motion, or *ascent*, of the air in the middle part of the column to prevent a vacuum forming near the top of the column. This ascending motion is the only possible way to replace the air which is leaving the column near the top. Figure 4.1(b) shows this situation schematically.

This is a very simple explanation of what can actually be a very complex set of processes. However, it goes a long way towards explaining why high pressure is usually associated with good weather whereas low pressure usually means bad weather. In a region of high atmospheric pressure the air is usually descending. As air descends it warms up, and if the air contains a certain amount of water vapour, which doesn't change as it descends, then because it is warming, the *relative humidity* of the air will decrease (see Chapter 2). If relative humidity is decreasing, clouds and therefore rain are less likely. In an area of low atmospheric pressure, the air is usually rising. As the air rises it cools, and at some point the vapour

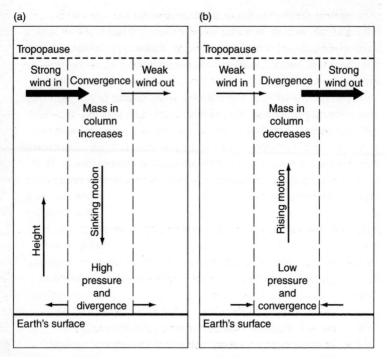

Figure 4.1 Schematic representations of the relationship between upper level convergence or divergence, surface pressure, surface winds and vertical air motions. (a) shows a situation with upper level convergence, high surface pressure, surface divergence and sinking motion. (b) shows upper level divergence, low surface pressure, surface convergence and rising motion.

in the air may start to condense, forming clouds. If the air then continues to rise, the clouds will get deeper and the cloud droplets will get larger and may get big enough to fall as rain.

Pressure maps

So far we've seen that the atmospheric pressure determines how the winds will blow near the surface of the Earth. This indicates whether there will be rising or sinking motion of air in the atmosphere and hence determines whether there will be clouds and rain, or clear skies. No wonder then that weather forecasters spend a lot of time looking at maps of atmospheric pressure, and that these maps are often presented as part of a weather forecast on television or the internet. Lines of constant atmospheric pressure called isobars are drawn on the map to reveal the 'topography' of the pressure field in the same way that height contours reveal the hills and valleys on a geographical map. Centres of high pressure are usually marked with an 'H', and low pressure centres (depressions) are usually marked with an 'L'. Regions of slack pressure gradients between weather systems are referred to as cols.

Here are some key points for interpreting these maps.

▶ *Air would like to flow from high towards low pressure. However, the rotation of the Earth means that the winds actually blow* parallel *to the isobars, not across them, with low pressure to the left in the northern hemisphere. This allows us to use pressure maps to assess the directions in which the air is moving.*
▶ *The spacing of the isobars indicates the theoretical strength of the wind. The closer the isobars (i.e. the larger the pressure gradient), the stronger the winds. The wind strength calculated from the spacing of the isobars is called the* geostrophic wind. *In reality the actual wind strength rarely corresponds exactly to the geostrophic speed, but looking at the isobar spacing works as a good general guide to the wind strength.*

The weather associated with areas of high or low pressure is not confined just to the centre of these regions. Areas of low pressure often have *troughs* extending some distance away from their centres whereas areas of high pressure often have *ridges* associated with them. The weather in these regions will be characteristic of the 'parent' weather system. See Figure 4.2 for an example pressure map.

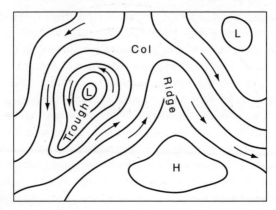

Figure 4.2 An example surface pressure map showing centres of low pressure (marked L), high pressure (marked H), a trough of low pressure, a ridge of high pressure and a col. Lines of constant pressure, called isobars, are shown, together with arrows indicating the directions of air motions around the map.

Let's now start looking at regions of the atmosphere which are characterized by either higher than normal or lower than normal atmospheric pressure. These particular regions, or *pressure systems*, will have characteristic weather associated with them due largely to the vertical motion going on within them.

High pressure: anticyclones

Having seen that high pressure is often accompanied by largely cloud-free, dry weather, we would expect weather systems with high pressure at their centres to be the 'good guys' in the battle between high and low pressure mentioned above. If we want an extended period of dry, sunny weather with warm daytime temperatures, at least in summer, then what we would like to see on the weather forecast is a map showing a region with high atmospheric pressure centred over our region of interest. In this section we will look in a bit more detail at the characteristic weather associated with areas of high pressure, or *anticyclones*.

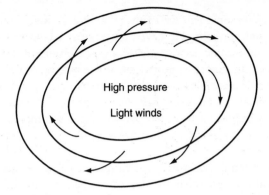

Figure 4.3 Isobars and surface wind patterns in an anticyclone in the northern hemisphere.

An anticyclone is an area of relatively high atmospheric pressure, associated with clockwise circulation of winds in the northern hemisphere (see Figure 4.3). There is no absolute value of pressure above which weather systems are classified as anticyclones; the central pressure just has to be higher than surrounding areas, usually with a closed isobar apparent on a weather map. The circulation of winds around an anticyclone can be understood in terms of the Coriolis effect. Due to the pressure gradient between the centre of an anticyclone and its surroundings, air wants to flow from the centre of the system where the pressure is highest, outwards towards where pressure is lower. The Coriolis effect will then kick in, turning this outward flow to the right so that the resulting pattern of winds is a circulation around the edge of the anticyclone, in a clockwise direction in the northern hemisphere, and anti-clockwise in the southern hemisphere.

As well as the clockwise circulation of the winds (in the northern hemisphere) anticyclones are also characterized, as we have seen, by *descending air*. However, particularly over the oceans, the descending air can trap a shallow layer of rather moist air near the surface which can lead to rather extensive and persistent layers of *low cloud*, perhaps only a few hundred metres deep or less, but enough to block out the Sun. People living on the coast of central and northern California for instance, or the west coast of South America, will be very familiar with this effect, particularly during their winter season. Near the centre of an anticyclone, winds tend to be very light or calm. However, move away from the centre and the wind can become deceptively strong, sometimes much more than one would expect from looking at the spacing of the isobars on a pressure chart. This is because winds around an anticyclone are what weather forecasters call *super-geostrophic*. This is particularly important to remember when trying to predict winds when looking at a pressure chart (see Chapter 9).

Anticyclones tend to be associated with extremes of temperature, both cold and warm. High pressure over a mid- to high-latitude continental area during winter can lead to very cold temperatures. Clear skies and long hours of darkness result in large amounts of energy radiating from the Earth's surface out to space. Light or calm winds mean that the air near the surface is not mixed up with the air above it, and so a thin layer of air in contact with the surface just gets colder and colder, remaining trapped below warmer air above. During the summer months the highest temperatures over land also tend to occur in anticyclonic conditions. In this case the clear skies lead to strong *heating* of the surface by the Sun. This in turn leads to warming of the air immediately above the surface. In such a case we might expect convection to set in, lifting the warm parcels of air up through the atmosphere so that they cool. However, the large-scale descent or *subsidence* of air within an anticyclone tends to form a strong temperature inversion at some height above the surface, which acts as a lid to prevent convection setting in.

AIR MASSES

Anticyclones tend to form in particular locations on the globe at different times of the year. Typical locations are over the subtropical oceans, most strongly during summer months, and over large mid- to high-latitude continental regions during winter. Because some anticyclones tend to form and stay in one location for extended periods, the air within them becomes more and more uniform over time, taking on the temperature and moisture characteristics typical of that location. So subtropical anticyclones over the oceans in summer tend to contain air that is warm and moist, whereas anticyclones over continental regions in winter tend to contain air that is uniformly cold and dry. These large, rather uniform regions are often referred to as air masses. Meteorologists give these *air masses* names that indicate where they have formed. Air masses which form on the polar side of the jet stream are called *polar* air masses, whereas those which form on the equatorward side of the jet are called *tropical* air masses, although *subtropical* would be a better name since they are typically formed between about 20 and 45 degrees of latitude in either hemisphere rather than deep within the tropics. Air masses are then further classified by whether they form over *continental* or *maritime* regions. Air masses are not confined to the region in which they form. Shifts in the position of the jet streams can lead to air making very large excursions from its source region. Tropical maritime air can be pulled a long way polewards in the warm sector of a mid-latitude depression (see below), and polar maritime air can make big equatorwards departures behind a cold front. Air that approaches the UK directly from the Arctic, an occurrence limited to the winter season, is referred to as *Arctic Maritime* (see Figure 4.4).

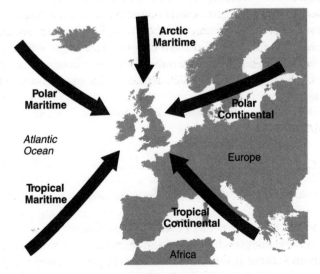

Figure 4.4 Source regions for air masses which affect the UK and western Europe.

As air masses move away from their area of origin they are modified slowly by the conditions of the surface over which they pass. Polar maritime air moving equatorwards over progressively warmer oceans will be warmed from below, whereas tropical maritime air moving towards the pole will be cooled in its lowest layers. This makes a big difference to the *stability* of the air, i.e. whether the air wants to rise or not. Air that is being warmed in its lowest layers will become more unstable – meaning that it is more likely to rise, producing convective cloud and showers. Air which is being cooled from below will be experiencing the opposite effect – cool air lying beneath warm air will not be able to rise due to buoyancy and so convective clouds cannot form.

Throughout the mid-latitudes, the type of air mass covering a region will have a large influence on the character of the weather. In some areas, changes in air mass type are pretty rare, so the weather can remain very similar for weeks on end. In other areas, changes in air mass occur on an almost daily basis as wind directions change and different weather systems pass through. It is in these areas where the weather is most changeable and least predictable. Coastal regions of the major land masses between about 40 and 60 degrees north and south sit at the intersections between both tropical and polar, and continental and maritime air masses. It is these locations that experience the most variability in the weather. The UK and the coastal fringes of western Europe, the US north-west, the Canadian west and east coasts, the coastal regions of north-east Asia, including Japan and, in the southern hemisphere, southern South America and New Zealand, all fall into this category.

Low pressure: mid-latitude depressions

Mid-latitude depressions or *cyclones* are the other basic currency of weather systems on this part of the globe. Entire books have been written on these systems and their accompanying *weather fronts* as they have such a major influence on the weather. Due to the rising motions of air associated with low surface pressure, depressions generally bring cloudy and often wet weather, so these systems are the 'bad guys' in that battle between high and low pressure that TV weather forecasters often describe.

Depressions are characterized by a centre of low pressure relative to the surrounding area, with at least one closed isobar on a weather map (see Figure 4.5). Again, there is no absolute value of pressure that defines a depression. A deep winter depression could have a central pressure as low as 910 millibars, whereas a summer system might have a central pressure in excess of 1010 millibars. Depressions are characterized by rising motion of the air within them, forming clouds and rain. Tightly packed isobars are a feature of many depressions, associated with strong winds. However, the winds associated with depressions are often much less forceful than might be thought from looking at the spacing of isobars on a weather map. This is because winds around a depression are *sub-geostrophic*, i.e. they are less than the wind speed that would be calculated from the spacing of the isobars. Even so, most instances of destructive winds in mid-latitudes are still associated with depressions.

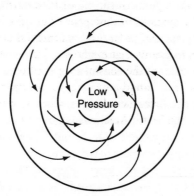

Low
Pressure

Figure 4.5 Isobars and surface wind patterns in a depression in the northern hemisphere.

Most depressions form near regions where the temperature changes rapidly in a short horizontal distance, known to meteorologists as *baroclinic zones*. Often these zones occur at the interface between warm tropical air masses and colder, polar air. There are several factors that might lead to a depression starting to form, and they are usually associated with features in jet streams. Places where the air flowing along the jet stream is slowing down, speeding up or changing direction are preferred locations for vertical motion of air through the depth of the troposphere beneath the level of the jet stream. In regions where that vertical motion is upwards, cloud will start to form as the rising air cools and the water vapour within it condenses. Such areas are known by forecasters as *development areas*. One consequence of ascent occurring within a column of air is that the surface pressure will start to fall, and another is that the air in the column will start to rotate. If that rotation coincides with a region of strong horizontal temperature gradient near the surface then the warm air will be pulled polewards and the cold air equatorwards by the winds, creating a 'wave' in the temperature gradient, and increasing the strength of that gradient. This development of a temperature wave adds to the development of the depression, as further ascending motion and falling surface pressure occurs in the region where the warm air is being drawn polewards. This region is ahead of the centre of the developing depression, and so as the depression moves towards it, the central pressure falls further. Thus there is a positive feedback process going on whereby motions in the upper atmosphere induce the development of a depression near the surface. The winds near the surface then lead to changes through the depth of the atmosphere that further reinforce the development of the depression.

Figure 4.6(a) shows a schematic representation of the development process at work. Figure 4.6(b) shows where in the atmosphere these development areas tend to be found. They often occur in association with upper tropospheric jet streams. The right-hand side of a jet 'entrance' region, where air is accelerating rapidly into the jet stream, is one preferred location, as is the left-hand side of a jet 'exit region' where air is decelerating as it flows out of the jet. Deviations in the direction of the jet stream known as *ridges* and *troughs* are also areas where upper level divergence may be found. A depression will often form initially in one development area, move along the jet without changing much and then deepen further as it passes through another development area.

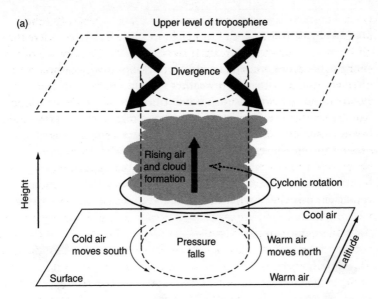

(a)

Upper level of troposphere

Divergence

Rising air and cloud formation

Cyclonic rotation

Height

Cool air

Cold air moves south

Pressure falls

Warm air moves north

Latitude

Surface

Warm air

(b)

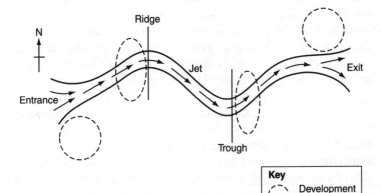

N

Ridge

Jet

Exit

Entrance

Trough

Key

Development area

Figure 4.6(a) schematic representation of how upper level divergence leads to ascent and falling surface pressure in a developing depression. (b) typical positions of development areas relative to an upper tropospheric jet stream. jet entrances and exits, ridges and troughs are locations where development areas often occur.

Frontal systems

While depressions are regions of generally unsettled weather, often embedded within such weather systems are lines of deep clouds associated with distinct bands of rather heavy rain. These bands are often the dominant features of a region of low pressure, bringing the most dramatic weather. They are known as *frontal systems* or often just as 'fronts'. In Chapter 8 we'll learn more about how fronts were discovered and how the terminology of fronts developed. Here we will look more closely at the weather conditions associated with this particular group of weather systems.

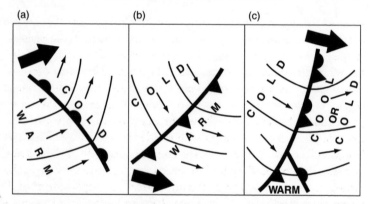

Figure 4.7 Representation of surface fronts on a weather map – (a) warm front, (b) cold front and (c) occluded front. Small arrows represent surface wind direction. Large arrow represents direction of movement of front. Thin lines are isobars.

Much of the rising air motion associated with depressions occurs in the vicinity of the boundary between the warm, moist tropical air and the colder polar air. These boundaries are known as *fronts*, with the *warm front* marking the line along which warm air is replacing cold air at the surface, and the *cold front* indicating the line along which cold air is replacing warm. On weather maps, the position of warm fronts is marked by a line with filled semicircles along the side of the line in which the front is moving. Cold fronts are indicated by a line with filled triangles indicating the direction of movement (see Figure 4.7).

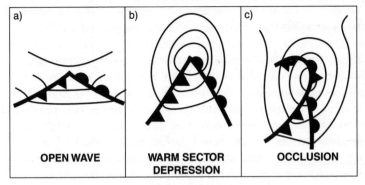

Figure 4.8 Sequence of events in a developing frontal wave depression.

Initially the warm and cold fronts in a developing depression form a shallow or 'open' wave pattern, but as the system develops and deepens, this wave becomes more pronounced. The region inside the wave formed by the warm and cold fronts is known as the *warm sector*, as it tends to contain warm, moist air of subtropical origin, in contrast to the cooler polar air masses outside the wave (see Figure 4.8). The air in the warm sector is not simply a static lump of warm air that moves with the depression. It is constantly being replaced by inflow from the equatorward side of the system, and removed by outflow both towards the pole and upwards. The airflow that performs this constant removal and replacement of the warm air is called a *warm conveyor belt*, and it is vital to the life of the depression (see Figure 4.9).

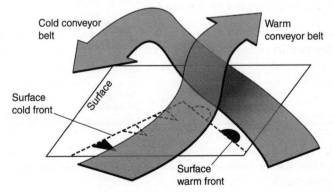

Figure 4.9 Schematic representation of the conveyor belts through a depression.

Without the constant throughput of warm moist air, the depression would soon run out of moisture to form clouds, and so the weather system would decay. The continuous feed of warm moist air into the system allows it to go through a life cycle that can last for several days rather than just a few hours. Ahead of the advancing depression, a flow of colder air wraps itself around the system, passing underneath the warm conveyor belt and then rising as it emerges on the poleward side of the system. This flow is known as the cold conveyor belt and it often leads to a distinct region of cloud developing poleward of the cloud band associated with the warm conveyor belt. This region of cloud is often referred to as a *cloud head* and can be a region of quite heavy precipitation.

As the weather system goes through its life cycle, the warm sector tends to get narrower as the cold polar air behind the depression wraps further around the system. Eventually, the warm moist air is lifted up off the surface, forming a wedge above the ground. The main air mass contrast at the surface is now between the cool, drier air ahead of the system, and the cold, polar air behind the system. At this stage the system is said to have started to *occlude* (which means 'to close up'), and the front is marked on weather maps by the occluded front symbol – a combination of the cold and warm front symbols (see Figure 4.10).

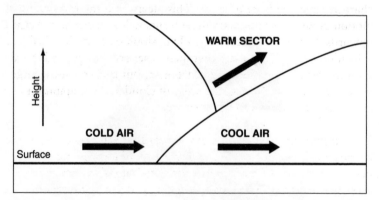

Figure 4.10 Vertical cross section through a depression as it begins to occlude and the warm air is lifted off the surface.

Be aware though that the occluded front symbol is also often used to mark bands of cloud and rain that haven't come about through this occlusion process. Very often the cloud head associated with the cold conveyor belt is also marked with an occluded front symbol.

Plate 2 shows a sequence of infrared satellite images of a developing depression over an 18-hour period. The band of very bright cloud on the first image has the very sharp polewards edge that is very typical of cloud associated with a strong jet stream. Six hours later this cloud band has developed a distinct 'S'-shaped curvature on its polewards edge, and a region of lower (greyer) cloud has started to appear on this flank. Over the next 12 hours this emerging cloud head becomes more distinct, with a slot of largely cloud-free air developing between it and the main cloud band. By the time of the final image, the system is becoming more organized in shape, looking less like an elongated band and more like a circular mass. The depression deepened (i.e. its central pressure *reduced*) by approximately 24 millibars during the period of these images.

Weather fronts aren't just surface features. They extend through the depth of the troposphere. They are rarely completely vertical boundaries between two different air masses, but instead have characteristic slopes with height. This means that the vertical motion of air associated with these fronts will also not be directly upwards, but will instead be a gradual, sloping ascent. Vertical velocities may only be a few centimetres per second, which is hard to measure with any standard instrument, but because of the large *mass* of air rising at fronts, the areas of cloud and precipitation associated with them can be extensive.

The warm front will always slope forwards ahead of the approaching weather system, with the air in the warm conveyor belt overrunning the colder, drier air ahead of the system. This gives rise to a typical progression of clouds that an observer on the Earth's surface would see as the front approaches. High clouds such as cirrus or cirrostratus would be the first to appear on the horizon. These will generally thicken and lower as the front approaches, being replaced by mid-level clouds such as altostratus and altocumulus (see Figure 4.11). Finally as the surface front passes over, thick layer clouds bring the heaviest precipitation. See Chapter 7 for more details of these cloud types.

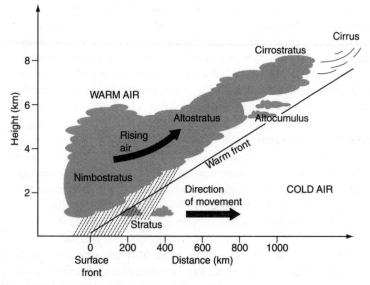

Figure 4.11 Schematic vertical section through a warm front.

The structure of cold fronts can be much more complicated than warm fronts. Whether the front slopes forwards or backwards, or a combination of both, depends on complex interactions between the air in the warm conveyor belt, the colder air behind the front, and very cold dry air which often descends from the upper troposphere behind the weather system. In perhaps the simplest case, the cold front slopes backwards from its position at the surface, with the colder air to the rear of the front undercutting the warm moist air in the warm sector, as shown in Figure 4.12(a). However, in many cases, cold dry air from the upper part of the troposphere behind the front actually *overruns* the warm sector. This creates a discontinuous frontal boundary and is usually referred to as a *split cold front*, where the upper section has pushed on ahead of the surface front. Figure 4.12(b) shows this type of development. One consequence of this splitting may be that there are two distinct bands of rain associated with the front instead of one. The rain from the upper section of the front would be falling in the warm sector, ahead of the position of the surface cold front. A conventional frontal analysis on a weather chart would just show the position of the surface cold front, and so the rain falling from the upper section of the front will apparently not be associated with a front at all. To account for this, forecasters sometimes use a special symbol on a weather chart to indicate the position of the upper section of the cold front. This symbol is similar to the standard cold front symbol, but with the triangles left unfilled.

Insight – Ana- and kata-fronts

At some fronts, the rain may be very light, or even non-existent, with the front being marked by just a band of cloud and a temperature change. An *inactive* front like this usually indicates that there is some large-scale downward motion of air above the front that is suppressing the development of clouds and precipitation. Such fronts are known as *kata-fronts*, from the Greek word meaning 'downwards'. An active frontal system, with deep layers of cloud extending up to the tropopause and heavy rain associated with it, is called an *ana-front*, from the Greek word for 'upwards'. These two cases are of course the opposite ends of a whole spectrum of possibilities, and most fronts fit in somewhere between. It is quite rare for both the warm and cold fronts associated with a depression to be equally vigorous. It is far more often the case that one of the fronts will be active, with deep cloud and heavy rain while the other is relatively weak.

FRONTS AND DEPRESSIONS IN THE REAL WORLD

It is important to remember that this description of depressions and their weather fronts is a generalization of an extremely complex and variable phenomenon. Every depression and front is slightly

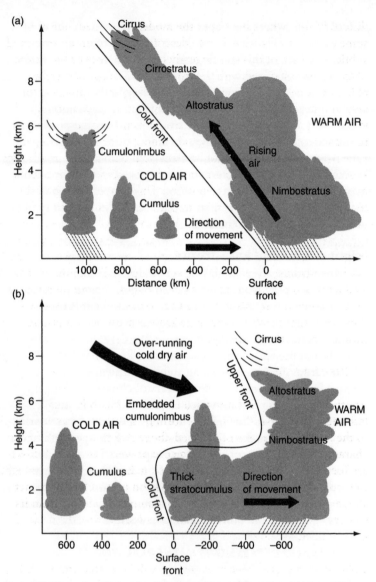

Figure 4.12 Schematic vertical section through a cold front – (a) a rearward sloping cold front, (b) a split cold front.

different, and although most fit the models presented here to some extent, it is not possible to describe all the permutations and subtleties which occur in real weather systems. One of the skills of the operational weather forecaster in mid-latitudes is to be able to recognize the unique features of every individual weather system that they encounter, and modify the *conceptual model* of the 'typical' depression that they carry around in their head to fit the actual system. Depressions and fronts also undergo changes as they move and develop. An initially active cold front may decay suddenly, or a rather weak band of rain ahead of the warm front may suddenly become more vigorous. This is particularly true in the summer months. Frontal systems during summer tend to be weaker than in winter because the equator-to-pole temperature gradient on which they form is also weaker. However, if a weak front moves across a land surface that has previously been warmed by the Sun, convection can break out along the line of the front, leading to a rapid intensification of the rain. Constant monitoring of the developing situation is essential to ensure that forecasters don't get caught out by assuming that a depression will just go through its 'textbook' life cycle.

CUT-OFF LOWS

Another type of depression which is quite common in mid-latitudes is the *cut-off low*. These form by a different mechanism to the frontal depressions described above and have very different characteristics. Instead of having a core of warm air, cut-off lows are formed when a region of cold air from high latitudes is brought towards the subtropics by a large deviation or meander in the jet stream. This region of cold air can be completely cut off from its source when the jet stream re-establishes itself across the neck of the meander, as shown in Figure 4.13.

Cold air sitting over a warmer surface will be heated from below and will become unstable, initiating convection. Thus instead of cloud and rain associated with fronts, the weather in cut-off lows is characterized by deep convective clouds and heavy showers. These can last for several days until the cold air is gradually 'warmed out'

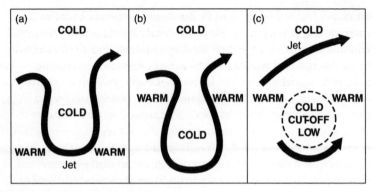

Figure 4.13 Formation of a cut-off low.

by the process of heating from below and convection that carries
that heat upwards through the troposphere. Such systems are more
common in the northern hemisphere where the jet stream tends to
be subject to larger deviations than in the southern hemisphere.
Cut-off lows are particularly common over the Mediterranean
region, where they can bring some very unsettled weather,
particularly during autumn and winter. The central pressure in
these depressions is usually not particularly low, so they don't
show up as particularly deep features on a surface pressure map.
However, on a satellite image the region of deep convective clouds
is a very clear indicator of the presence of a cut-off low.

POLAR LOWS

Another distinctive type of depression, which is associated with
cold air moving over a warmer surface, is the *polar low*. Once
again, these systems tend to be most common in the northern
hemisphere and form exclusively over the oceans. In the North
Pacific they form at quite low latitudes (40–50 degrees north),
whereas in the Atlantic they tend to be more genuinely polar,
forming at latitudes north of 60 degrees north. This is because
polar lows tend to form over strong ocean surface temperature
gradients, and the different nature of the two ocean basins mean
that the strongest temperature gradients in the Atlantic are
further north than they are in the Pacific. Polar lows are rather

small in size, perhaps just a few hundred kilometres in diameter compared to the typical scale of a frontal depression of a thousand kilometres. Polar lows lack frontal systems, but are characterized by convection as the cold air in the core of the low is heated from below as the system moves over warmer waters. Often the convective clouds become organized into a 'comma' shape making a very distinctive pattern in satellite pictures (see plate 3).

Precipitation from polar lows tends to fall almost exclusively as snow or other types of frozen precipitation such as hail or ice pellets. Polar lows derive their energy from a warm ocean surface so tend to die out quite quickly on making landfall. However, they can still produce significant snowfalls over land before they do collapse, and some of the heaviest snowfalls over the UK and Ireland are brought by polar lows. Systems that track down the North Sea can give heavy snowfalls down the east coast and across East Anglia. Occasionally a polar low will run down the Irish Sea, bringing snow to the Welsh coast, and even on rare occasions across the south-west peninsula. Here the snow can be enhanced over the high ground of Exmoor and Dartmoor, and strong winds that accompany the snow can bring blizzard conditions and deep snowdrifts.

Tropical weather systems

Across the tropical regions of the globe, weather systems are often very different from the systems seen at higher latitudes. Contrasts in atmospheric pressure in the tropics tend to be much smaller than in mid-latitudes, so systems are not so much characterized by their atmospheric pressure. That said, some tropical storms and hurricanes sometimes have very low surface pressure indeed associated with them and in some ways these systems are similar to mid-latitude depressions. However, because the Earth's surface in the tropics is very warm, the nature of weather systems here is generally quite different to the systems seen in mid-latitudes. In this section we will look at the reasons for these differences and examine in detail some of the weather systems that are unique to the tropical atmosphere.

In the tropics, much of the motion of the atmosphere is a direct result of the strong surface heating by the Sun. Air near the surface gets warmed by contact with the surface, becomes less dense and hence more buoyant and so starts rising. As it rises it cools until at some point condensation of the water vapour in the air starts to occur. This condensation forms clouds and also releases latent heat into the air, making it warmer and more buoyant again and so the rising process continues, forming more cloud and eventually rain. Of course this process, known as *convection*, can occur anywhere on the globe if air near the surface becomes warmer than its surroundings. What makes it such a major feature of the tropics? To start with, the heating of the Earth's surface by the Sun in the tropics is very strong. But equally important is that, because the surface and the air above it is much warmer in the tropics, values of *absolute* humidity in the atmosphere can also be much higher. A kilogram of air near the Earth's surface in the tropics might contain 30 grams of water vapour, whereas in mid-latitudes perhaps only ten grams in each kilogram of air would be made up of water vapour. This means that there's a lot more latent energy locked up in the tropical air, waiting to be released as the air rises.

This type of *convective* motion might just form individual clouds, with a horizontal size of about a kilometre and a height of perhaps 5–10 kilometres. In the tropics however, convection often becomes *organized* into larger systems, with many individual ascending plumes of air embedded within them. There are several complex processes going on in the tropics that help to cause these bigger systems. Plate 4 is a satellite image of central and southern Africa, showing bright convective clouds on a wide range of scales, from individual clouds a few kilometres across up to systems with diameters of several hundred kilometres. There are many different mechanisms by which convection in the tropics becomes organized into larger-scale weather systems, and we'll take a look at a few of these now.

Wind convergence

Regions where the low-level winds are *converging* can be preferred locations for organized convective activity. *Convergence* of winds just means that the winds are carrying air into a region more quickly than they are taking air out of the region. As air converges into a region, simple physics says that the air must go somewhere – it can't just keep accumulating at one location. The outlet is vertical motion, and since the presence of the solid earth beneath it stops the air going down, it must go up (see Figure 4.14).

This process can enhance the *convective* ascent, creating large areas where clouds and rain are more likely to occur. Tropical surface winds also tend to be laden with plenty of water vapour, and so a region of wind convergence will generally also be an area of *moisture convergence*, providing the fuel which will go on to form clouds and rain in a developing weather system. The Inter-Tropical Convergence Zone (ITCZ) (see Chapter 3) is a very large region in which we would expect to see convergence of the low-level winds. On any given day that convergence is stronger in some parts of the ITCZ and weaker in others, accounting for the clumpy rather than continuous nature of the clouds in the ITCZ when seen from satellite.

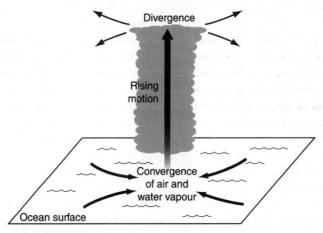

Figure 4.14 Schematic representation of how convergence of surface winds causes vertical motion and cloud development.

African easterly waves

One rather distinctive tropical weather system that features wind convergence is called the *African Easterly Wave*. In some ways this wave is similar to the waves that form in mid-latitudes on the strong temperature gradient there. However, this system forms over North Africa in the *reversed* temperature gradient between the hot Sahara Desert and the rather cooler land and ocean surface to the south. Because the temperature is *decreasing* towards the north, the wave runs *westward*, across the Atlantic and towards the Caribbean. In the part of the wave where the surface winds are convergent, convection often starts to form, and as this passes over the warm waters of the tropical Atlantic there is a ready source of heat and moisture to keep the convection going. Surface pressure falls in the centre of the convective area and so the winds start to converge even more towards the centre. Add some turning of these inrushing winds due to the Coriolis effect and the system starts to rotate. The rotating, inrushing winds cause lots of evaporation of the warm ocean surface waters beneath them, providing more and more fuel for the system. The vertical motion and surface convergence increase, the system spins faster and we have a tropical cyclone. Many of the hurricanes that hit the Caribbean or the south-eastern US start out this way. While only a very small percentage of African easterly waves end up forming hurricanes, their appearance on satellite imagery is closely monitored by the various storm forecasting and warning centres. Hurricane Dean, which hit Jamaica on the 20 August 2007 and then moved on to make landfall on the Yucatan peninsula of Mexico (see plate 5), started life as an area of convective activity just off the coast of Africa ten days earlier.

Tropical cyclones

Tropical cyclones, the rotating storms of which hurricanes are a subgroup, can form in ocean basins other than the North Atlantic of course. The Pacific experiences cyclones to the north of the equator on both the western and eastern side of the basin, and to the south of the equator on the western side. The Indian Ocean also sees cyclones both to the north and south of the equator, with cyclones occurring in the summer season in each hemisphere. Tropical cyclones in the South Atlantic had never been observed until March 2004. Then a system with all the appearance of a hurricane hit the coast of Brazil, causing quite extensive damage. Whether this was a complete freak, an extremely rare event or a sign of things to come in the future is not yet clear. However, it does serve as a warning that the atmosphere can sometimes throw up some pretty unexpected conditions.

There are several ingredients for the formation of tropical cyclones. The main ones are listed here.

▶ *All tropical cyclones form over oceans, as they need a source of moisture to maintain their clouds and act as an energy source. It could be said that tropical cyclones are fuelled by drawing energy out of the ocean beneath them.*
▶ *Sea temperatures need to be above about 26.5°C (80 F). This helps to explain why we never see cyclones south of the equator in the East Pacific where the ocean is just too cold. Not only does the ocean surface need to be very warm, but the water in a reasonably deep layer below the surface (at least 50 metres) also needs to be very warm. The strong winds in a cyclone generate a lot of ocean waves. These in turn cause a lot of mixing between the water at the surface of the oceans with water lower down. If the sub-surface water is too cold, as it gets brought to the surface it will lower the temperature at the ocean surface thus reducing the energy supply to the developing storm.*
▶ *Cyclones never form within five degrees of latitude of the equator. Close to the equator the Coriolis effect is very small and so rotation cannot develop.*
▶ *Cyclones require the middle part of the troposphere to be moist. If the currents of rising moist air within the developing storm encounter dry air as they ascend, these* updraughts *will be diluted as they mix with the drier air, thus losing their energy.*

▶ *Tropical cyclones only form in regions of* low vertical windshear. *This means that the wind speed and direction in which the storm is forming must not change too much with height above the ground. If the windshear is too strong, and in particular if the upper-level winds are very strong, then the clouds forming at the top of the developing storm will be blown away before the storm can become fully established through the depth of the troposphere. The development of an El Niño in the Pacific Ocean generally leads to upper-level* westerly *winds across the tropical Atlantic that oppose the low-level easterly trade winds in this region. This enhances the vertical windshear leading to fewer tropical cyclones in the Atlantic during El Niño years. La Niña has the opposite effect. During 1997, a year with a large El Niño event, there were only seven named storms in the Atlantic, with only three of these reaching hurricane strength. The following year saw a strong La Niña event, and there were 14 named storms in the Atlantic, of which ten made it to hurricane strength.*

Tropical cyclones generally track from east to west as they move in the easterly trade winds which prevail throughout the tropical ocean basins. Cyclones in the West Pacific and Atlantic often take on a more northwards component as they develop. In some cases this means that storms can make landfall at latitudes that we wouldn't normally class as tropical, with Japan and the eastern seaboard of the US being particularly in the firing line. Cyclones usually weaken fairly rapidly after making landfall, as their main energy source of a warm ocean surface is removed. Some storms however do manage to maintain enough of their identity to reinvigorate if they subsequently move back over an ocean. Storms in the Caribbean have been known to pass across Central America into the Pacific. While the storm generally loses much of its intensity as it passes across land, there is still enough structure remaining to be reactivated as the storm arrives once again over warm ocean waters.

Insight – 'The eye of the storm' and 'the calm before the storm'

Tropical cyclones differ from mid-latitude depressions in many respects. One obvious one is that tropical cyclones form completely within a tropical air mass, so there are no frontal boundaries associated with them. Instead, tropical cyclones often appear to be almost circular. The heaviest rain is usually located near the centre of the system, but as the intensity increases, the heaviest rain and thickest cloud often form into a set of spiral arms, which appear to spiral in a clockwise direction out from the centre in the northern hemisphere, while the system as a whole rotates anti-clockwise. Another distinctive feature of the tropical cyclone as it increases in strength is the formation of a clear 'eye' at the centre. In this region the air is descending and the winds at the surface are very light. People who have been directly beneath the passage of the eye of a tropical cyclone often report an eerie quiet and strange quality of light in this region, before the eye moves away and the winds increase rapidly again.

Prior to the arrival of a tropical storm, the weather can be absolutely perfect – clear blue skies and calm winds giving no hint of what's to come. This 'calm before the storm' is partly due to the air around the edges of a cyclone sinking downwards to compensate for all the rising motion going on within the storm. This sinking motion suppresses any clouds. Often the first sign of the approaching storm is an increase in ocean wave heights with no apparent wind. The falling surface pressure in the developing storm, which may still be several hundred kilometres away, generates ocean swells which spread rapidly away from the storm centre.

There is a lot more information on tropical cyclones and hurricanes in Chapter 6, with details on their devastating effects when they make landfall and information on how they are tracked and forecast.

The Madden-Julian Oscillation

Mention the Madden-Julian Oscillation (known by meteorologists as the MJO) in conversation and you'll probably be met by blank looks. Unlike El Niño, which has worked its way into the public consciousness, the MJO is pretty well unheard of outside the meteorological community. However, it has a big effect on the weather throughout much of the tropics and even has impacts at higher latitudes. The oscillation is basically a switch between very disturbed convective conditions with heavy rain and strong winds, and a quieter period of clear skies and light winds. This swing from disturbed weather to quiet and back again has a period of about 60 days but is somewhat unpredictable – it can go through several cycles in a row and can then disappear for months on end.

The disturbed, or *active*, phase of the MJO starts off in the equatorial Indian Ocean as convection gradually builds up and becomes organized on a scale of several thousand kilometres. This phase of the MJO is shown as a satellite picture in plate 6(a). This area of extensive cloud and rain then starts to move eastwards, towards Indonesia and Malaysia. It is followed by a period of strong westerly winds, and then clearer skies and lighter winds. Plate 6(b) shows the situation ten days after the first picture with the western Indian Ocean now largely clear of cloud apart from two tropical storms either side of the equator on the very western side of the Indian Ocean. As the area of disturbed weather moves across the islands of Indonesia it can bring torrential rain, flooding and mudslides to the islands before moving out into the West Pacific. On reaching the Date Line, the convection usually dies away as it moves over the colder waters of the East Pacific. However, a disturbance in the upper-level winds often continues to move eastwards like a wave in the atmosphere, and on reaching the warmer waters off Central America and the Caribbean, the convection can often flare up again. In fact, many of the tropical storms that form in the East Pacific do so during the active phase of the MJO. The upper level wind disturbance continues eastwards around the globe and on arriving back over the Indian Ocean, the active convection flares up again, starting a new cycle of the MJO. A full cycle of the MJO takes between one and three months, and once a cycle is underway, its behaviour is generally quite predictable. Forecasters in Indonesia watching the clouds and rain build up in the western Indian Ocean can be fairly sure that those disturbed conditions will be reaching them in perhaps one to two weeks time. Sometimes however, the MJO dies away completely for several months, and if this happens then it is not possible to predict with any certainty when the next cycle will begin.

Summary

In this chapter we have looked at some of the systems that bring different parts of the world their typical weather. In forecasting terms these are the things that meteorologists pay most attention to as it is these systems that cause day-to-day variations in the weather and make the atmosphere more or less predictable. However, knowing what type of weather systems are likely to affect your area of interest, and even being able to forecast the tracks and development of these systems is not the end of the story in terms of predicting what type of weather is going to occur. Local geography can have a large impact on the weather conditions caused by these weather systems. Two locations separated by only a few tens of kilometres may experience completely different weather during the passage of a depression over the area because they are either side of a range of hills. It is this type of geographical variation which puts the fine detail onto the large-scale weather patterns and we will investigate this further in the next chapter.

10 THINGS TO REMEMBER

1 *A map of atmospheric pressure can tell us which direction
   the wind will be blowing, how strong the wind will be and
   where the air will be rising or sinking. This makes pressure
   maps an invaluable tool to weather forecasters.*

2 High pressure *weather systems are usually associated with
   settled weather, although they are not always sunny. Some of
   the world's major high pressure systems are often cloudy.*

3 *In mid-latitudes the day-to-day variability of the weather is
   determined by the origin of the* air mass *that is influencing
   the region.*

4 Low pressure *weather systems or* depressions *are usually
   associated with unsettled weather with strong winds and the
   formation of clouds and rain.*

5 *The formation of depressions is determined by acceleration
   and deceleration of winds in the jet stream, leading to
   preferred* development areas *for these weather systems.*

6 *Mid-latitude depressions are characterized by the presence
   of warm and cold fronts – bands of rising air leading to the
   formation of clouds and rain due to the contrast between
   warm and cold air masses within the depression.*

7 *Every depression is slightly different, so the nature of the
   frontal systems and the weather associated with them varies
   considerably. This means that most fronts do not conform to
   the 'textbook' picture.*

8 *In the Tropics, many weather systems are driven by* converging
   low level winds. *This convergence leads to rising motion,
   generating clouds and rain.*

9 *Tropical cyclones always form over warm tropical oceans that provide the heat and water vapour necessary to fuel an intense weather system.*

10 *You may never have heard of the* Madden-Julian Oscillation *but it has a strong influence on tropical weather patterns on timescales of weeks to months, making tropical weather more predictable on these timescales.*

5

Local weather effects

In this chapter you will learn:
- *how local geography can modify the weather*
- *about the effects of mountains, coasts and the nature of the land surface on creating localized weather conditions*
- *how large cities can create their own climate.*

One thing that we soon learn from experience is that any given weather system is unique – it produces slightly different conditions from any previous example of the same type. Another thing that starts to become clear when we look at weather systems in more detail is that they can have quite different effects on different locations. The weather associated with a particular system can vary considerably, even over quite short distances. This is because the weather can be modified considerably by the underlying surface of the Earth. Some places experience unique weather conditions due to very particular interactions between large-scale weather systems and the local geography. In this chapter we will investigate some of the ways in which particular aspects of geography can lead to distinct local weather effects. Some of these will be unique to a particular location, such as the Mistral wind of the Rhône Valley in central southern France, whereas others will occur more generally in a particular type of terrain, such as sea breezes around coastal regions or the urban heat island effect in major cities.

Local weather effects can be broadly grouped under three headings:

▶ The effects of topography *(hills, mountains, valleys, etc.)*
▶ The effects of the underlying surface *(forest, grassland, urban, etc.)*
▶ The effects of proximity to the ocean or other large body of water.

In many places, the local weather conditions will turn out to be due to a combination of two or even all three of these factors. Scale is also a consideration. Some local weather effects are very local indeed – a small valley or hollow can experience night minimum temperatures several degrees cooler than places only a few hundred metres away. On an even smaller scale, have you ever noticed that on a still clear summer evening, the air can feel noticeably warmer if you just walk underneath a tree? On a much larger scale, major mountain ranges such as the Rockies in North America and the Alps in Europe can exert an effect on the weather for many tens or even hundreds of kilometres downwind. Both ranges produce winds that blow often enough over a wide enough area to have acquired their own well-known names – the Chinook over the Rockies and the Föhn over the Alps.

Topographic effects

Mountains and even rather modest hills can have a big influence on the air that flows over and around them. There are many ways in which the weather that occurs on or near hills and mountains is different from what would be experienced if those hills were replaced by a flat plain. It is important to remember that, while many hills may not be very tall by comparison with say the Himalayas, the atmosphere itself isn't very deep either. The troposphere, the layer of atmosphere in which most of the weather that we experience at the surface is generated, is on average only about ten kilometres deep. A hill of only 500 metres (roughly 1,500 feet) is occupying the bottom five per cent of the troposphere, so will still exert quite an influence on the air around and above it. Ranges such as the Himalayas, and to a slightly lesser extent the Andes, the Rockies and the Alps, are actually occupying a considerable proportion of the depth of the troposphere. In the case of the Himalayas particularly, the mountains are so high that they act effectively as a dam to the air in the troposphere, preventing airflow from one side to the other and leading to completely different weather conditions on either side of the range. During the Indian Summer Monsoon, the southern flank of the Himalayas experiences torrential rain, while only 100 kilometres (60 miles) to the north, the Tibetan plateau remains largely desert-like.

Perhaps the most obvious effect of hills and mountains is on temperature. In general, the higher you go, the colder it gets. On average, the temperature decreases by about 6–7°C (11–13 F) for every one kilometre gain in altitude. There can of course be exceptions to this rule. For instance, under a strong area of high pressure, or after a calm clear night, a situation can develop where the temperature actually increases with height, sometimes quite markedly over a few tens of metres. In hilly regions, this type of temperature inversion is most obvious when a layer of mist or fog is trapped in the valleys, with the tops of the hills emerging into clear skies. In general though it is safe to assume that it will be colder as altitude increases. Hill walkers setting off from a warm, sunny valley should always be aware of this and carry extra clothing to deal with the drop in temperature that even a modest gain in height will bring.

Things start to get more interesting when winds start to blow across hills and mountains. In the case of a completely isolated hill, the air could either pass over or around the hill, depending on several factors. In the case of a long ridge of hills or mountains such as the Andes in South America or the Rockies in North America, the air is forced to pass over the top, or through any gaps in the ridge. When air rises, it expands as it experiences lower pressure. The expansion is accompanied by cooling. This cooling occurs at a very predictable rate: 9.8°C (18 F) for every kilometre that the air rises as long as the air remains unsaturated. This rate of temperature is referred to as the *Dry Adiabatic Lapse Rate* (DALR). As the air cools, it may reach its dewpoint temperature, at which point the water vapour in the air will start to condense and clouds will form. Once condensation occurs, the air cools at a much slower rate as the condensation itself releases heat which is absorbed by the air. If the air is very close to being saturated before it even starts to rise then the hills or mountains don't need to be very high for clouds to form. This will come as no surprise to anyone who has spent time on the windward coast of any country or island with even moderately sized hills. The west coasts of Britain and Ireland are often shrouded in cloud as moist air from the Atlantic is forced to rise over hills which, on the scale of the world's great mountain ranges, are hardly very significant. At the same time, the eastern sides of these countries can be sitting under clear skies. The air rising over the hills has a high water vapour content as it has been travelling across the Atlantic Ocean prior to arrival on the west coast of Britain or Ireland. As the air rises over the coastal hills it cools, and soon reaches saturation. Cloud forms, and some of the water may actually fall out as rain or drizzle. On passing down the far side of the hills, the air is now drier, having lost some of its water content. The descending air warms, and soon becomes unsaturated, so there are fewer clouds on this side of the hills.

Insight – The Indian Western Ghats – the wettest hills on Earth?

The west coast of India, which is subjected to an onshore south-westerly wind from the Indian Ocean during the monsoon season, is among the wettest places on Earth from June to September. The air is forced to rise by the chain of hills known as the Western Ghats which extend down the west coast. Rainfall totals during the monsoon season amount to well over a metre (40 inches) along the length of this mountain chain. The height of the hills is pretty modest – between about 500 and 1,000 metres (1,500–3,000 feet) but the uplift they cause is sufficient to generate massive amounts of cloud and rain. These hills experience almost no rain at all for several months of the year, but medieval hill forts built

(Contd)

There is one particular mechanism whereby even quite modest
hills can enhance the rainfall amounts associated with a weather
system such as a warm front, particularly if those hills happen to
face a coastline. As the front approaches, the frontal clouds may
be several kilometres above the tops of the hills. However, if the
air near the surface is saturated, low stratus-type clouds will form
as the air flows up the hills. These clouds by themselves may not
be able to produce much more than light drizzle. However, the
raindrops falling from the frontal cloud above the hills collide with
the cloud droplets in the lower cloud layer, and will often combine
with these drops. Thus the raindrops that reach the ground are
larger than they would have been if they had fallen through clear
air and so the rain is heavier. This mechanism is called the *seeder-
feeder* effect, with the frontal clouds above the hills known as the
seeder clouds and the low stratus being the *feeder cloud* (see
Figure 5.1).

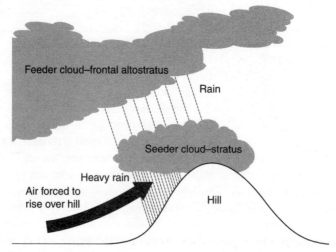

*Figure 5.1 The seeder-feeder effect leading to increased rainfall on the windward side of a range
of hills.*

Just as interesting as what happens to air as it rises over hills and mountains is what happens to the air as it then descends on the other side of the ridge. As air descends, it is compressed as it experiences higher atmospheric pressure, and so it warms. If the air became saturated due to the cooling as it rose, then as it warms on the way back down again it may reach a temperature at which it is no longer saturated. The cloud base on this side of the hill will occur at this level. If some rain fell from the clouds as the air rose over the hills, then the air passing down the lee slope will be drier than it was on the windward slope and so the cloud base will be higher. Perhaps more significant is that the air will warm up *more* on the way down than it cooled down on the way up. The net result of this effect is that when the air reaches the foot of the hills on the lee side it can be several degrees warmer than it was at the foot of the hills on the windward side (see Figure 5.2).

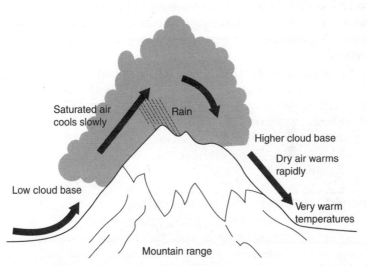

Figure 5.2 The Föhn effect, showing air cooling as it rises over a mountain range, producing cloud and rain. As the air then descends on the lee side it warms rapidly.

Meteorologists call this the *Föhn effect* after the warm wind that blows down the lee side of the European Alps, often raising the temperature by 10°C (18 F) in a few hours, and causing rapid snowmelt during spring. This effect is not limited to the Alps of course. In the North American Rockies, the same effect is called a *Chinook* wind, from the native American words for 'snow eater'. Such an effect is also possible over hills much smaller than the Alps or the Rockies. In a strong westerly flow during summer, the east coast of Scotland can be several degrees warmer than the west thanks to the Föhn effect and, if conditions are right, Inverness and Aberdeen can be the warmest places in Britain.

Air that sits over a high plateau, especially during winter, can cool, becoming rather dense. Dense air will try to flow downhill, resulting in strong, cold, downslope winds. The extreme example of this phenomenon occurs around the edges of Antarctica. The air over the plateau of central Antarctica becomes very cold indeed, particularly during winter when there is no sunlight for six months. Temperatures of –40°C (–40 F) are not unusual. In the centre of the Antarctic land mass, winds are light, but around the edges, where there are very steep slopes, the cold air can rush down the slopes at speeds exceeding gale force. The strongest winds occur in the steep-sided valleys that run down from the plateau to the coast. Unfortunately, the Antarctic research base at Mawson is located right at the foot of one such valley and so is subject to some of the most vicious winds on Earth. A less dramatic example of the same effect occurs along the Adriatic coast of Slovenia and Croatia. Cold air that has built up over the mountainous plateaus of Slovenia or Bosnia flows down over the Croatian coast in a wind known locally as the *Bora*.

Insight – Mistral – the 'master' wind

Particular pressure patterns in the vicinity of hills and mountains can lead to some very characteristic winds. High pressure extending over central France during winter and spring leads to a build-up of cold air over the Massif Central. Because of these hills and the mountains of the Alps, the cold air flowing down the slopes will be strongly funnelled along the Rhône Valley, emerging on the Mediterranean coast as a strong wind known locally as the *Mistral* (see Figure 5.3). During the winter this wind will be very cold but in summer, when it occurs less often, it can be very hot. In both cases it often extends out over the Mediterranean Sea for several hundred kilometres, resulting in pretty rough conditions for shipping in the area. Locals along the Rhône valley and the coast around Marseille claim that the onset of the Mistral is preceded by a feeling combining irritability and nervous energy. When the wind itself starts blowing, many claim to suffer from headaches. The name Mistral means 'master', with the implication that the wind dictates people's behaviour on the 100 or so days per year when it blows.

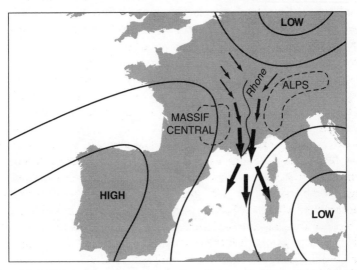

Figure 5.3 The Mistral wind over southern France, funnelling down the Rhône valley between the Massif Central and the Alps.

In general, steep-sided valleys will always act to funnel, and often speed up, the wind. In the Kali Gandaki Valley of Nepal, which runs through the Himalayas from Tibet down to the southern foothills, the wind direction is usually only ever straight up or straight down the valley. The valley is considered to be the deepest valley in the world, flanked as it is on either side by two mountains over 8,000 metres (26,000 feet) in height. The direction of the wind above the valley has very little bearing on the wind within the valley itself, which is more a result of the daytime heating and cooling of the slopes around the valley. During the night the cold air from the elevated plateau at the head of the valley flows down the valley, and is joined by cold air flowing down the slopes on either flank – see Figure 5.4(a). During the day this direction reverses as warm air from the lower reaches flows up the valley, as shown in Figure 5.4(b). This type of pattern is quite common in valleys in high mountains, but in the Kali Gandaki it is very marked, and the up-valley winds during the afternoon can be very strong, regularly reaching 30 mph. Natives of the area can tell the time of day simply from the strength of the wind and whether it is blowing up or down the valley. These winds in the Kali Gandaki valley blow on about 300 days of the year regardless of what direction the wind is blowing outside the valley.

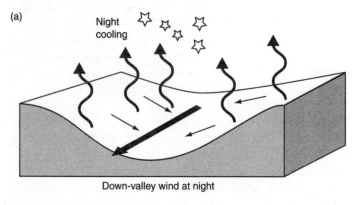

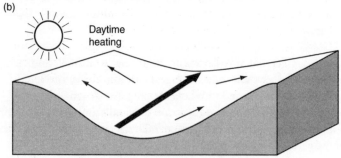

Figure 5.4 Valley winds – (a) shows the night-time situation, with cold air draining down the valley, (b) shows the daytime situation with warm air flowing up the valley.

Downslope winds can be very turbulent and gusty. The airflow across ridges can, in particular conditions of atmospheric stability, actually detach from the hillside on the downwind side, resulting in less friction acting to slow the wind down. At some point downwind of the summit, the wind will hit the surface again, resulting in very strong gusts that, if the large-scale wind direction and speed remain steady, can affect the same place for several hours. These conditions are not uncommon in cities such as Denver on the eastern side of the Rockies. Even modest hills can lead to damaging gusts in downslope winds.

Insight – The Sheffield storm, February 1962

On the 16 February 1962, the city of Sheffield, to the east of the English Pennines, suffered considerable wind damage during a windstorm that lasted several hours. Damage was concentrated in an area in the centre of the city, with about 60 per cent of the buildings within this area suffering some structural damage. Cranes on building sites collapsed and one of the 45-metre (150-foot) floodlight pylons at the Sheffield United Football ground was brought down. Three people were killed, all by falling chimneys and whole rows of prefabricated houses, built to replace housing destroyed during World War II, were flattened. Average wind speeds were measured at over 70 mph, with gusts much stronger than this. The particular conditions that led to this damage were very specific. A very strong wind was blowing from west to east across most of northern Britain, but above the Pennines there was a very strong *temperature inversion*, i.e. the temperature was increasing with height. This meant that the air that was being forced to rise over the Pennines was then being forced back down towards the ground very rapidly on the far side. Where this rapidly descending air was reaching the ground coincided exactly with the centre of Sheffield, and the situation remained this way for many hours.

In a stable atmosphere, waves forming in the airflow on the lee side of mountains are a common occurrence. Air that is forced to rise over the hills then starts to descend again on the lee side. However, if conditions are right, the air starts to oscillate about its original altitude as it flows along with the wind, forming waves. These waves can extend for several hundred kilometres downwind of the hills that caused them, as shown in Figure 5.5. Glider pilots are big fans of these *mountain waves*.

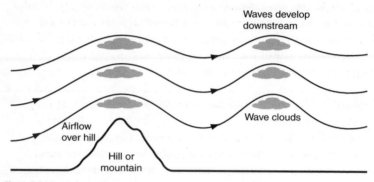

Figure 5.5 Mountain waves occurring downstream of a range of hills or mountains in stable atmospheric conditions.

The upward section of the wave provides a reliable region of rising air that the pilot can use to gain altitude before setting off on a cross-country leg of a flight. In satellite pictures, the waves are often revealed by alternate lines of cloudy and clear air *parallel* to the hills over which the wind is blowing. The clouds form in the crests of the waves while the troughs remain cloud free. These bands remain stationary relative to the ridge. From the ground, these *wave clouds* often resemble almonds or lenses with smooth, curved upper surfaces, hence their scientific name of *lenticular* (lens-like) clouds. Isolated wave clouds can resemble flying saucer shapes, and may be responsible for several reported UFO sightings every year! Plate 7 shows an example of this type of cloud over the Alps.

Surface effects

We know that the climate of a particular region can affect the vegetation in that region. In a warm, wet and humid tropical region the plants respond to the heat and plentiful water supply, resulting in the growth of luxuriant rainforest. In the deserts beneath the descending branch of the Hadley circulation in the subtropics, rain is scarce and temperatures range from searingly hot during the day to rather chilly at night. Few plants grow beyond a few tough shrubs. Clearly the climate controls the natural vegetation. What we tend to forget is that the process works the other way too. The nature of the Earth's surface and its vegetation (or lack of it) can also affect and modify the weather. Over many hundreds or even thousands of years, man has gradually modified the natural vegetation in many parts of the world, and this has undoubtedly had an effect on the weather in these parts of the world. However, when the vegetation is changed suddenly, perhaps over only a few years, we can see and even measure the impact. In regions of tropical rainforest where the forest has been cleared to make way for agriculture, or simply cut down for the timber, there are measurable effects on the humidity and temperature, with humidity decreasing and temperature generally increasing. In regions where very large areas of forest have been cleared, rainfall amounts have also been observed to decrease.

Insight – The urban heat island

Another big influence of man on the Earth's surface over the last century or so is the development of cities, with ever larger areas being covered by buildings and concrete. Perhaps the main effect of this development is on temperature. Building materials and concrete tend to heat up strongly during the day as they absorb sunlight. A road surface for instance can get much hotter than an adjacent area of grass or even bare soil. This heat is then released gradually to the atmosphere through the evening and night, resulting in higher temperatures. Overnight temperatures in the centre of a large urban area can be 5°C (9 F) or more warmer than the surrounding countryside. This effect is known as the urban heat island, and is particularly noticeable in calm or light wind conditions when there is little movement of air between cities and rural areas. As winds increase, the air in the city starts to be replaced by cooler air from the surrounding area and so the effect becomes weaker.

Urban areas can also have an effect on the winds. The presence of large buildings acts to generally slow down the wind, but also introduces more turbulence and gustiness. Narrow streets flanked by high buildings can also act to funnel the wind in the right conditions, which will actually speed up the wind. Sometimes, again in the right conditions, the increased turbulence of the winds and the additional heating of the surface in cities can act to trigger convection more rapidly and strongly than in the surrounding area. This can lead to storms with heavy showers and perhaps even thunder that may either sit over the city or move away into the countryside downwind.

It is not only cities that have the effect of slowing down the wind. In general, winds will always be lighter over a land surface than over open water due to increased friction, and the taller and denser the vegetation, the stronger this effect will be. Within a dense forest, the winds will always be quite light or even calm at the surface, no matter how strong the wind above the trees is. There are some rules of thumb which give an idea how much the wind will be slowed by underlying surfaces. In general, a land surface will always slow the wind down by between 25 per cent and 50 per cent from what it would be over an open ocean surface. Cities and forests will have the largest effect, whereas low vegetation and open fields or grassland will have a smaller impact on the strength of the wind. Friction over land also always has the effect of *backing* the wind from its original direction. For example, a weather observing station on the west coast of an island may be recording a westerly wind. However, another observing station a few miles inland will be recording a wind which is blowing from the *south-west*, and is weaker than at the coast, due to the effect of friction over the intervening land surface. In any kind of storm conditions, whether a tropical hurricane or a mid-latitude depression, wind speeds will always be stronger over exposed coasts than over inland areas. During the English 'Great Storm' of 1987 (see Chapter 6 for more details), the strongest winds were recorded at Worthing on the Sussex coast and at Gorleston on the Norfolk coast.

Man has also influenced convection in the countryside by replacing natural vegetation with cultivated crops. Plants such as trees and shrubs have fairly deep root systems that can access water in the ground to a depth of several metres. During dry conditions, the plants draw water up through their roots, which then passes up through the stems and out into the atmosphere through the leaves, increasing the local humidity. If these plants are replaced by shallow-rooted crops, the deep soil water is no longer accessible and the humidity above the crop decreases. The drier air results in fewer convective clouds and less precipitation. This of course has the effect of making the soil drier still, thus amplifying the original effect. We can sometimes see this effect very clearly in satellite pictures or aerial photographs of regions where cultivated land sits next to areas of natural vegetation. Convective clouds occur over the natural vegetation but not over the crops. One place where this effect is often seen is along the *rabbit-proof fence* in the Australian outback. The fence marks the boundary between cultivated fields and the natural vegetation. It stretches for thousands of miles. Ironically it doesn't actually keep out rabbits at all, but does appear to keep out convective clouds and showers, often resulting in a marked line between clear skies and cumulus clouds – see colour plate 8. Removal of the natural vegetation to make way for grazing in the Sahel region of northern Africa is also thought to be a contributing factor to the extended drought that has affected that area intermittently since the mid-1970s.

The replacement of natural vegetation by crops usually means that at certain times of the year, the soil is left bare. In mid-latitude regions this generally occurs in the winter season, which is usually wet, so that the soil never dries out. However, in the semi-arid tropics and subtropics, the crop-free period usually corresponds to the dry season and so the soil becomes very dry and dusty. Any occurrence of strong winds at this time can pick up the dust and loft it into the atmosphere, reducing visibility and making conditions generally unpleasant. Dust storms are common throughout the arid regions of Asia, parts of Australia and particularly Africa. African dust storms originating over Niger,

Mali, Senegal and Mauritania often result in dust being carried out over the Atlantic Ocean by the easterly trade winds. Dust originating over North Africa is regularly deposited in the islands of the Caribbean and dust plumes extending from the African west coast can often by seen in satellite pictures of the Atlantic. See plate 13 for an example.

Tiny, microscopic particles from the surface are present in the atmosphere all the time and can affect the weather in more subtle ways than simply reducing visibility in a dust storm. These particles, known as *aerosol*, also act as sites for water vapour to condense onto, forming cloud or fog droplets. Over the oceans, these particles tend to consist largely of salt from ocean spray. The particles can be quite large and are relatively few in number. Over land, aerosol particles come from a number of different sources such as wind-blown dust, soot from natural and man-made fires, sulphur particles from volcanic eruptions and various man-made pollutants. It is not perhaps immediately obvious how the different numbers of aerosol particles in the air might affect the weather. However, there are two main effects of the difference between land and oceans. The first is on the visibility. Over land regions, particularly during periods of high pressure when the aerosol particles in the air near the surface are trapped there by the large-scale *descending* motion of the air above, the large number of very small particles can lead to haze, significantly reducing the visibility. Although such haze is unlikely to be as thick as fog, it can cause problems for pilots of light aircraft navigating by visual contact with the ground. In particular, haze can affect the *slantwise visibility* from an aircraft making it hard to see the runway as the aircraft makes its approach to an airfield. Haze particles can also act as *nuclei* for fog droplets, providing sites for water vapour to condense onto as the temperature drops on a still night.

Insight – London 'pea-souper fogs' and modern 'smog'

The 'pea-souper fogs' of London prior to the 1950s are a classic example of artificially enhanced fog, where pollution particles from coal fires led to much thicker fog than otherwise might have been expected. On some nights buses had to drive at walking pace through the city preceded by a guide on foot as visibility was reduced to less than ten metres (33 feet). The Clean Air Acts passed by the UK Parliament in the 1950s put paid to this particular menace, but in today's modern cities, visibility can still be reduced by pollution, with exhaust emissions from motor vehicles taking the place of smoke from coal fires. The smog (a mixture of smoke and fog) of cities such as Los Angeles, Beijing and Athens are all due to the large numbers of motor vehicles on the roads combined with large-scale weather conditions which prevent the pollution escaping from the lowest layer of the atmosphere. Such smog can have a detrimental effect on public health as well as having an impact on the weather. Incidents of prolonged smog usually coincide with increased hospital admissions for patients with breathing problems.

The second effect of the number of aerosol particles in the air is on the formation of showers. Over the oceans, there are fewer aerosol particles in the air than over land. As clouds start to form, the water vapour starts condensing onto the available aerosol particles. Over the oceans, the water vapour is spread between the relatively few aerosol particles and so the cloud droplets become large quite quickly. Over land the same amount of water is spread between many more particles so the individual droplets are much smaller. The result is that the cloud droplets over the ocean can grow big enough to fall out of the cloud as rain showers much more quickly than over the land. Hence showers can fall over the oceans and near to coasts from clouds that are much less deep than over inland regions. Ragged cumulus clouds over the oceans that look fairly innocuous can still produce quite vigorous showers, and small hailstones can often form in oceanic convective clouds of quite limited depth.

Coastal effects

The effects of being near to an ocean have already been mentioned several times in the sections above. It is sometimes difficult to separate coastal effects from the impacts of mountains or the changes in the nature of the land surface. For instance, if a warm, moisture-laden wind is blowing towards a coastal region, then even a small amount of uplift due to hills on land can lead to the formation of a blanket of low, stratus cloud that clings to the coast. However, if the land was completely flat, this cloud may not form.

One phenomenon that is clearly associated with coastal regions is the sea breeze. On a typical summer day, as cumulus clouds start to build up over the land, the formation of a *sea breeze* can often be responsible for keeping the coastal fringes free of cloud, which is good for holidaymakers on the beaches. However, the sea breeze generally brings in cooler air from the ocean which isn't quite such good news.

The reason for the formation of a sea breeze is the different rates at which land and ocean are heated up by sunshine. Land surfaces are warmed rapidly by sunshine, and the air immediately above them also warms through contact with the warm surface. As a result of this warming, the column of air above the land expands, as shown in Figures 5.6(1) and (2). The ocean surface warms much less quickly than land. The heat from the Sun penetrates beneath the surface of the water and so the same amount of heat is spread through a layer of water several metres deep. Because of this, the air above the ocean surface hardly warms at all. The contrast in temperature between the air over the land and over the ocean sets up a wind at some height *above* the surface that flows *from the land towards* the ocean. This wind above the surface is balanced by a wind *at the surface* that flows from the ocean towards the land as shown in Figure 5.6(3) and (4).

This is the sea breeze. At first this wind will be limited to the immediate coastal fringe, but as the day goes on it can penetrate further and further inland. In tropical regions where the Sun heats the land very strongly, sea breezes can extend well over 100 kilometres inland during the day, whereas in more temperate latitudes where the heating effect of the Sun is weaker, the sea breeze is unlikely to extend more than a few tens of kilometres inland from the coasts. Sea breezes always bring cooler air over

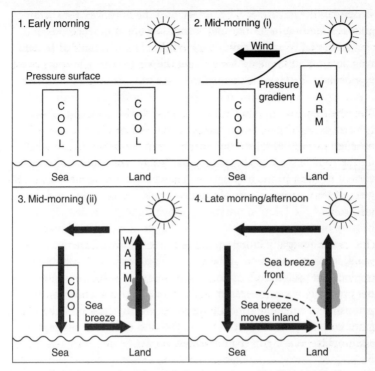

Figure 5.6 The development of a sea breeze during a warm summer day. The initial wind is offshore above the surface as the air over land warms more than air over sea. The sea breeze develops as a return circulation and then penetrates inland during the day.

the land, counteracting the heating effect of the Sun which triggers convective clouds. Hence there can often be a cloud-free strip inland from the coast as cumulus clouds build up further inland. The *sea breeze front* is defined as the boundary between the cool oceanic air and the warm inland air. This front never has the large-scale cloud and rain that we would associate with a true frontal zone, but it can be a region of shower activity as the cold sea air forces the land air to rise as it moves inland, triggering convection sooner and more vigorously than would happen if the sea breeze wasn't present. This effect can be particularly strong over a peninsula where two sea breezes can often penetrate inland from different directions. Where the two sea breeze fronts meet will be an area of strong *convergence* (see Chapter 4), and the only way for the air to go is up, leading to more rapid formation of cumulus clouds and showers than would occur if the heating of the surface by the Sun was the only mechanism for causing the air to rise. The south-west peninsula of England is one place where this phenomenon of *sea breeze convergence* is often observed. The north and south coasts will often be free of cloud while a line of showers extends down the centre of the peninsula, further enhanced by the presence of the high ground of Dartmoor and Bodmin Moor. Holidaymakers on the beaches can be enjoying a lovely day while those who have chosen to walk in the hills can be having a rather wet and miserable time. Plate 9 shows a pair of satellite images, one visible and one infrared, of the UK and northern France on a spring day when the heating of the land is setting off convective clouds and showers. The clouds are generally lining up along the direction of the near-surface wind, but there are many other complex patterns developing due to the local geography. Clear zones just inland of some coasts indicate the presence of sea breezes, and a line of cloud along the south-west peninsula of England appears to have been triggered by some small feature on the north coast of Cornwall, and enhanced perhaps by a sea breeze moving in from the south coast.

At night-time, the conditions which led to the formation of a sea breeze are reversed. The land surface cools quite rapidly once the Sun goes down. The ocean surface temperature meanwhile changes very little. This results in a wind that blows offshore, called, not surprisingly, a *land breeze*. Land breezes are generally somewhat weaker than their daytime counterpart, but they can be enhanced if the coast is hilly or mountainous. Cold air currents that drain down the hills at night add to the land breeze and can carry cold air many miles out to sea. This effect can be seen on a fairly regular basis around the islands of Indonesia and Papua New Guinea. During the afternoon and early evening, convective cloud and showers are located over the islands. These die out overnight, to be replaced by convective cloud and showers over the ocean channels between the islands in the early morning. In this case, land breezes from two or more islands can lead to convergence over the sea, further enhancing the shower activity.

Sea breezes are largely a summertime phenomenon, when strong heating by the Sun warms the land more than the sea. In the winter, it is often the case that the ocean can be warmer than adjacent land regions. If cold air moves over a relatively warm ocean, then the air is heated from below, and showers may start to form. However, if that same cold air moves over a cold land surface, the atmosphere becomes more stable and showers are unlikely to form over the land. In these conditions, somewhere like the British Isles can be surrounded by convective cloud and showers over the oceans, while over land the skies can be perfectly clear. However, if there is an onshore wind, some of the showers may be blown over land. These showers will eventually die out as the cold land beneath them cuts off their heat supply, but in a strong wind, the showers may penetrate many kilometres inland before they finally die out. Plate 10 is a satellite picture of the UK in a northerly airflow during winter. Most of England and Wales is free from cloud, but convective clouds (and probably showers) are running across East Anglia, blown in by the wind from the North Sea.

Topography can play a part in this process too. Hills or mountains can block the showers from running very far inland, but over a flat plain, there is nothing to stop them. A north-westerly wind over Britain will often lead to showers from the Irish Sea, sometimes of snow, running across Liverpool and the Cheshire Plain and penetrating all the way to Birmingham, over 100 kilometres inland. However, if the wind swings around to a northerly, the showers from the Irish Sea will run into the hills of North Wales and will not get very far inland.

An *alongshore wind* can also lead to a convergence zone along a coast due to the effect of friction over the land on the wind direction, as shown in Figure 5.7.

Consider a north-facing coast. If the wind offshore is westerly then friction over the land will result in the wind direction over the land being closer to south-westerly. This means that there will be a convergence line along, or very close to, the coast. In the right conditions this can lead to a line of showers which constantly regenerate in the same place. This type of mechanism was partly

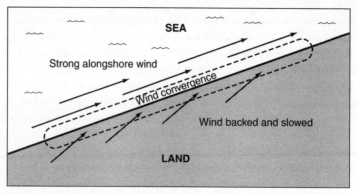

Figure 5.7 Coastal convergence due to friction over the land causing the wind to back. This can lead to the development of a stationary line of showers parallel, and very close to the coast.

responsible for the rain that brought flash floods to the Cornish village of Boscastle on the 16 August 2004 (see Chapter 6 for more details). Of course, if the wind is blowing in the opposite direction then the convergence line will be replaced by a region of *divergence*, in which showers would be rather unlikely to form.

No coastline is perfectly straight for mile after mile, and the occurrence of bays, headlands and changes in the orientation of the coast make the simple pictures described above rather more complicated. Elevated headlands can lead to local wind convergence if an alongshore wind blows around rather than over the headland (see Figure 5.8).

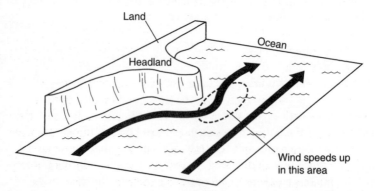

Figure 5.8 The presence of an elevated headland can lead to the speeding up of alongshore winds as they flow around the headland.

Sometimes this can be just enough to trigger convection and shower formation. This type of flow will also result in a speeding up of the wind around the headland, an effect that coastal sailors need to be aware of (see Figure 5.8). Sailors who are familiar with a particular stretch of coast will start to learn what to expect in different wind directions, and can anticipate changes in wind strength and direction. Serious competitive sailors who are going to be racing in unfamiliar waters, such as Olympic sailing teams, will often employ their own meteorologists. These people will record and study the winds in the race area over a period of many weeks prior to the event. They will try to quantify and map all the local variations in wind strength and direction that might just give their team the edge when racing begins.

Fog

Fog is a localized weather condition that can depend on the presence of hills and valleys, the nature of the land surface and proximity to bodies of water and so spans all three sections of this chapter. Fog forms when the air adjacent to the surface becomes saturated and condensation occurs. This can be achieved in two ways. Firstly the air can be cooled to its dewpoint temperature, and secondly water vapour can be added to the air to bring it to saturation without changing its temperature.

How does 'radiation fog' form?

The most common way for fog to form over land regions is through *radiative cooling* of the Earth's surface on a clear, calm night. The layers of air in contact with the surface also cool, and if cooled to their *dewpoint* then fog will form. If the wind remains absolutely calm then this fog will hug the ground, resulting in rather ethereal shallow fog, perhaps only one metre deep. A very little light wind will help to mix the cooler air up through the

(Contd)

atmosphere, deepening the fog layer. This type of fog is known as radiation fog. Terrain and land surface will both influence where fog forms. At night the coolest air will drain into valley bottoms and hollows, so these will be the first places where fog appears. The amount of moisture in the upper layers of soil or lying on the surface will also influence fog formation. Fog will form most readily where there is an available water source. Hence lakes, marshes and even small ponds or streams may be the first places for fog to form. Fog is also far more likely after a day of rain than after a dry day as there will be standing water to feed the air with water vapour as it evaporates. Low-lying land is doubly likely to see fog since not only is it likely to experience cooler temperatures on a calm, clear night, but is also likely to have damper soil as water drains downhill. Radiation fog is by its nature very localized, particularly as it is just starting to form, and it is this patchiness that makes it such a hazard to motorists.

The other main way of forming fog involves the movement of warm, moist air over a progressively cooler surface. The lowest layers of air may be cooled to their dewpoint, when condensation will occur and fog will form. This type of fog is called *advection fog*, and it is most likely to occur over oceans. Typically advection fog forms as warm, subtropical air moves polewards over cooler oceans. Fog around the southern and western coasts of the UK in spring and early summer forms through this mechanism. The coasts of California are also subject to advection fog as air from the Pacific moves over the rather cold coastal waters. Clearly some wind is needed for advection fog to form, but if the wind is too strong then turbulent mixing may lift the fog up into a layer of stratus cloud instead. Plate 11 is a satellite image showing an extensive area of advection fog in the North Sea between Britain and Denmark. The winds are from the east, blowing the fog onto the east coast of Britain where it is penetrating quite some distance inland in places. This type of sea fog on the east coast is known locally as *haar* or *fret*.

Using local knowledge

As this chapter has shown, there are many different ways in which local geographical conditions can affect the weather that you experience on a day-to-day basis. The main effects of locality are summarized below.

► **Coastal regions** – *These tend to be cooler, windier, cloudier and wetter than inland areas particularly if the prevailing wind direction is from the sea.*
► **Hills and mountains** – *Upland areas tend to be cooler, windier, cloudier and wetter than lowland areas. So a combination of hills in a coastal region can lead to particularly cloudy, wet and windy conditions.*
► **Urban areas** – *Large urban areas can have a noticeable effect on the weather. Temperatures are generally warmer, especially at night, and winds, while often being lighter than in countryside areas, can often be gusty, and due to funnelling along streets and around high buildings, the wind can be rather strong in very specific locations.*

By having a good understanding of how the local geography can modify the weather conditions, we can start to use our knowledge to add detail to a weather forecast which may not fully take account of localized effects. In Chapter 9 we will start to think about exactly how we can do this in a way that 'adds value' to the forecasts provided through the media by national weather services and other weather forecasting agencies.

10 THINGS TO REMEMBER

1 The nature of the Earth's surface can considerably modify the weather conditions that affect any given location.

2 The presence of coastlines, mountains, large bodies of water and changes in vegetation all act to affect local weather conditions.

3 Mountains force air to rise over them. As the air rises it cools which can generate clouds and even rain. Even rather modest ranges of hills can enhance rainfall amounts from frontal systems through the 'seeder-feeder' mechanism. Rain falling from frontal clouds above the top of the hills falls through lower stratus cloud generated by air flowing over the hills. The raindrops coalesce with the cloud droplets in the low cloud making them bigger and hence making the rain heavier.

4 The Föhn Effect is another phenomenon caused by air flowing over mountains. Air descending on the lee side of a mountain range such as the Rockies or the Alps can sometimes be several degrees warmer than on the windward side.

5 Urban areas are often warmer than the surrounding countryside – an effect known as the 'urban heat island'. On calm clear winter nights, city centres can be 5 degrees or more warmer than nearby rural locations.

6 The rougher the Earth's surface the more the wind is slowed down and 'backed' in direction by the effects of friction. In a dense forest a strong wind at treetop level can be reduced to near calm at the surface.

7 A haze of tiny particles of pollution from urban areas can act to reduce visibility. In extreme cases these pollution particles can act as sites for fog particles to form, resulting in smog.

Plate 1 Satellite image showing cloud patterns in the ITCZ.

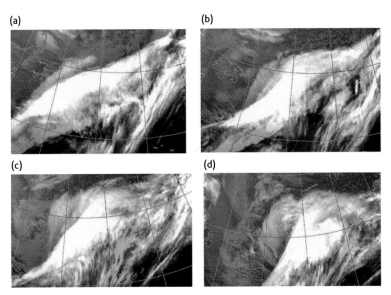

(a)

(b)

(c)

(d)

Plate 2 A sequence of satellite images, six hours apart, showing the rapid development of a deep depression from an initial cloud band associated with the jet stream, through the initial stages of wave development, the appearance of an emerging cloud head and finally reaching maturity.

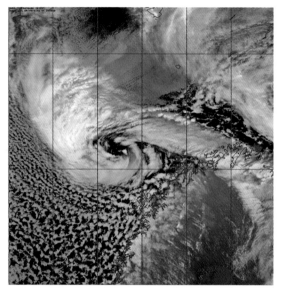

Plate 3 Satellite image of a polar low near the coast of northern Norway.

Plate 4 Satellite image of central and southern Africa at 12 UTC, showing convective systems on a wide range of scales from individual thunderstorms to larger areas of organized convection.

Plate 5 Satellite image of Hurricane Dean approaching Mexico on 20th August 2007.

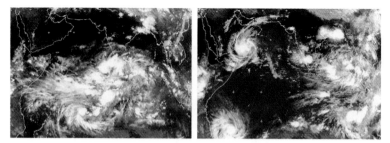

Plate 6 Two satellite images of the Indian Ocean, one week apart, during May 2002. In the first image the cloudiness associated with the active phase of the MJO is centred over the Indian Ocean. In the second image this cloud has moved eastwards over Indonesia, leaving the Indian Ocean largely clear, apart from a pair of tropical storms in the west of the basin.

Plate 7 Wave clouds forming over the Alps.

Plate 8 Aerial photograph looking along the rabbit-proof fence in western Australia. Shallow cumulus clouds are confined to the uncultivated side of the fence.

(a)

Plate 9 (a) Infrared and
(b) visible satellite
images of convective
clouds over the UK
during late spring,
when the land is
warmer than the
ocean. A line of deep
cloud, associated
with heavy showers,
can be seen down
the western side of
the country.

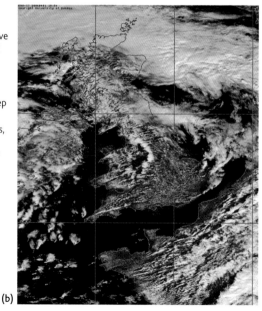

(b)

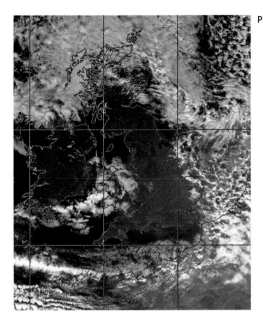

Plate 10 Satellite image of convection over the ocean around the UK during the winter when the land surface is colder than the ocean.

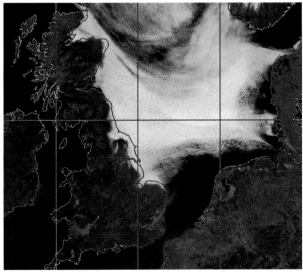

Plate 11 Satellite image showing extensive sea fog in the North Sea, running onto the east coast of the UK.

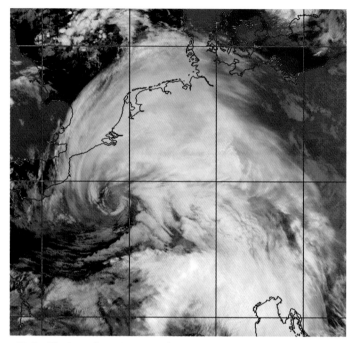

Plate 12 Satellite image for 0754 UTC on the 26 December 1999, showing the 'Lothar' storm that brought widespread damage to parts of Europe.

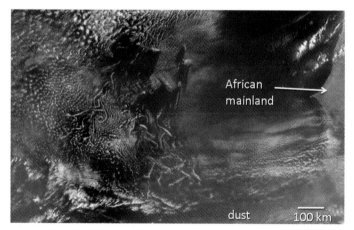

Plate 13 Satellite image of a dust-plume from West Africa crossing the Atlantic Ocean. The dust shows up as a dirty brown colour over the dark ocean surface.

Plate 14 A single-cell cumulonimbus cloud.

Plate 15 A multi-cell thunderstorm, clearly showing a tilt with height. New convective clouds are developing on the side nearest the camera.

Plate 16 Altocumulus cloud associated with an approaching warm front. At this stage the Sun can still be seen through the cloud, but 30 minutes later the cloud had thickened and steady rain was falling.

Plate 17 High, thin altocumulus cloud not associated with any frontal feature.

Plate 18 Tufts of high cirrus clouds. The lower grey clouds have been formed by cumulus clouds flattening out in the evening as the Earth's surface starts to cool.

Plate 19 Rainfall radar image of the UK at 2100 UTC on 19 June 2007. The bright colours indicate very heavy rain, and parts of north-east England were experiencing flash flooding at this time.

Plate 20 Visible satellite image from a geostationary satellite.

Plate 21 Infrared image from a geostationary satellite, for the same time as the visible image in Plate 20.

Plate 22 Water vapour satellite image from a geostationary satellite for the same time as the images in Plates 20 and 21.

Plate 23 A developing cumulus cloud. At this stage there is no ice apparent in the top of the cloud and no rain is falling at the base, but a shower isn't far off.

Plate 24 Cumulus cloud showing some flattening, which could be due to building high pressure or an approaching front.

Plate 25 The same cumulus cloud 15 minutes later, showing a thin layer of milky white cirrostratus above the cumulus. A warm front is on its way.

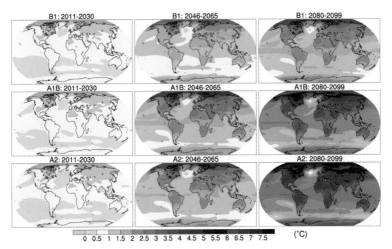

Plate 26 Multi-model mean of annual mean surface warming (surface air temperature change in °C) for the scenarios B1 (top) A1B (middle) and A2 (bottom), and three different time periods, 2011–30 (left), 2046–62 (middle) and 2080–99 (right). Anomalies are relative to the average of the period 1980 to 1999.

Plate 27 'Red sky at night...' In this case the red sky was probably due to increased levels of atmospheric pollution.

8 Sea breezes form due to the temperature contrast between a warm land surface and a cool ocean surface as the land heats up during the day. At night the land cools down and the direction of the sea breeze reverses, becoming a land breeze which blows offshore.

9 Radiation fog forms as air is cooled to its dewpoint, usually by radiation to space on calm clear nights. This type of fog will form first in low lying areas or in locations near a body of water or very damp soil.

10 Advection fog forms as warm air is blown with the wind over a cool, wet surface such as an ocean or large lake. The coastal fogs of California and the haar and fret on the East coast of the UK are types of advection fog.

6

....

Wild weather

In this chapter you will learn:
- *the importance of forecasting extreme weather events*
- *which particular weather conditions can be described as extreme*
- *why extreme weather in one location might be considered quite ordinary somewhere else.*

In most inhabited parts of the globe, the weather that occurs on a day-to-day basis only has a minor impact on daily life. Months or perhaps even years can go by without anything happening that could be described as severe weather. But when weather events occur that really do have a big impact, including loss of life, we'd like to know about them in advance. These are the situations in which forecasters really earn their money. Get it wrong and people will remember for a long time. Just ask Michael Fish, the Met Office forecaster who presented the BBC TV weather forecast on the 15 October 1987. He will never be allowed to forget his words: 'We had a phone call at the weather centre from a lady who had heard that there was a hurricane on the way. Well don't worry, there isn't.' Fish followed up that statement by saying that the weather *would* be very windy, but that part of his forecast always gets edited out in replays. That night a storm hit the south-east of England which caused extensive damage to trees, buildings, power lines and vehicles. Although technically it wasn't a hurricane, it was one of the most intense storms to affect that part of the country for many years. A government enquiry was launched to discover why the event hadn't been well forecast and the name of Michael Fish will always be associated with that storm despite the fact that he was only presenting a forecast produced by a national weather service.

There's something of a morbid fascination with weather that wreaks destruction and even endangers human life. Perhaps it's a reaction to the awesome power of the atmosphere and the realization that we and our environment are pretty fragile. Among weather enthusiasts this fascination can be taken to extremes. Tornado chasers like to try to get as close to a twister as possible and a hurricane making landfall will always be greeted by a few cameramen trying to capture the destructive power of the winds and waves on film. Extreme weather of this sort will always be big news. Journalists always like to give a label to such events – 'killer tornadoes', 'Great Storm', etc. – and the phrase 'freak weather' seems to be used for just about any event that is slightly out of the ordinary. Global warming is often trotted out by the media as a possible cause for damaging weather, despite the fact that experts would never try to attribute a single weather event to climate change. Forecasting extreme weather is always more difficult than forecasting run-of-the mill conditions, but as forecasting models improve and the tools available to forecasters become more sophisticated there's no doubt that forecasts are getting better. In this chapter we'll look at some of the weather types that are considered to be extreme, examine how they form and look at how difficult they are to forecast.

Intense mid-latitude depressions

The 'Great Storm' of October 1987 mentioned above was an unusually intense depression system of the sort described in Chapter 4. Wind speeds around the coasts of southern Britain and northern France touched hurricane strength, and even in inland areas of southern England winds reached gale force, which is a pretty rare event for this part of the country. Sixteen people were killed in southern England and the death toll may have been much higher if the storm had passed through during the day rather than overnight. Another intense depression crossed Belgium, northern France and Germany on Boxing Day 1999. Named 'Lothar' by the German weather service, this was a small depression in terms of size, but it packed winds well in excess of gale force, which again is rare for inland areas of continental Europe. About 100 people lost their lives and there was widespread disruption to power supplies for many days afterwards. The satellite image in Plate 12 shows this storm moving across northern Europe. The strongest winds are located beneath the tight swirl of cloud over northern France. Britain and north-west Europe are affected by perhaps 100 depressions in a typical year, so what makes some depressions reach this kind of intensity, and can they be forecast far enough in advance to give some useful warning to people in their path?

The first thing to remember is that gale-force winds are often associated with mid-latitude depression systems. However, these systems typically follow a track that takes the strongest winds across the oceans, or rather sparsely populated areas such as the west coasts of Scotland and Norway, Iceland and the Faroe Isles for Atlantic storms, and the Aleutian Isles, Alaska and the Canadian north-west in the Pacific. They only tend to grab our attention when they take a track across more densely populated areas. In the case of the 1987 storm, there was considerable damage caused in London. Perhaps it would be cynical to suggest that it became such a big news story because that's where most of the newspapers and TV broadcasting companies have their headquarters, but with millions of trees blown down and power supplies to much of London and the south-east of England brought down it certainly was an event which brought major disruption to the lives of millions of people.

There are several factors that can lead to a depression developing destructive power of this sort. All mid-latitude depressions form on a temperature gradient between warm tropical air and cold polar air, and if this gradient is particularly strong then there is more potential energy for the developing depression to feed on. One location where depressions often intensify rapidly is off the east coast of the USA and Canada during the winter season. The temperature contrast between tropical and polar air masses is enhanced by the contrast in surface temperature between the cold continental land mass and the warmer ocean temperatures in the Atlantic. A similar situation exists on the western side of the Pacific, where warmer waters near Japan contrast strongly with the cold Asian land mass. Storms that form or pass over these areas often undergo very rapid intensification, changing from rather insignificant systems to very intense storms packing gales and heavy rain in periods of less than 24 hours. These so-called *coastal storms* can often lead to heavy falls of rain or snow in the coastal regions and very strong winds out to sea. One example of such a storm was the so-called 'QE2 storm' in September 1978 that inflicted considerable damage on the ocean liner Queen Elizabeth II off the eastern seaboard of the USA. The intensity of the 1987 'Great Storm' was also due in part to the temperature contrast between very warm tropical air being drawn into the centre of the system and the rather colder polar air surrounding it.

As discussed in Chapter 4, it is usually developments high up in the troposphere that lead to the formation of depressions. A very strong jet stream in the upper troposphere can be a contributing factor to the formation of very intense depressions. A strong jet in itself is not enough however to guarantee the formation of intense depressions. What is also needed is strong acceleration or deceleration of the wind *within* the jet stream. Small changes in the direction of a strong jet stream known as *short wave troughs*, or small regions of extremely strong winds embedded within the jet stream, known as *jet streaks*, can lead to particularly vigorous rising motion of air through the depth of the troposphere. Once this rising motion is established, condensation will start to occur, leading to cloud formation, heating of the mid-troposphere and further ascent. All this will be accompanied by rapid falls in surface pressure and the intensification of surface winds (see Chapter 4).

The advent of frequent, high-quality satellite imagery has certainly improved the forecasting of these rapidly deepening systems. They tend to have a very distinctive pattern of clouds when viewed from space, with what forecasters call a *cloud head* emerging from a band of frontal cloud with a very sharp polewards edge. There is also often a clear slot or 'dry intrusion' apparent between these two main areas of cloud. This indicates that air from near the tropopause, or even from the stratosphere is being drawn down into the centre of the system that can further enhance the development of the system. These days forecasters are getting very good at recognizing such features, but in 1987 the quality and frequency of satellite images available to the operational forecasters was rather poor compared to the present day, and so the 'Great Storm' was not identified during its developing phase. Figure 6.1 shows a schematic representation of the cloud patterns associated with a rapidly developing low. The sequence of satellite images in plate 2 shows a real example of such a system.

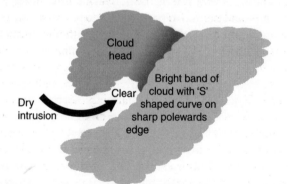

Figure 6.1 Sketch of the cloud head and dry intrusion features that often indicate the presence of a rapidly developing, intense depression.

Monitoring of satellite imagery by forecasters is perhaps the key to accurate forecasts of intense depressions. Computer weather forecasting models will often underestimate the strength of this type of depression meaning that forecasters have to be alert to the signs that indicate that something unusual is happening. Forecasts of the Lothar storm of 1999 were greatly improved by the input of human forecasters watching the satellite imagery and comparing it with the forecasts produced by computer. See Chapter 8 for more details of the forecasts for this particular storm.

Hurricanes

It goes pretty much without saying that all hurricanes are extreme events. The *minimum* sustained wind speed for a tropical storm to be classified as a hurricane (or *typhoon* in the West Pacific) is 119 km/h (74 mph). That's 57 km/h (35 mph) *higher* than the definition of 'gale force' on the Beaufort Scale (see Chapter 7), so clearly even the weakest category of hurricane is going to cause damage to buildings, trees and power lines wherever it hits land. But strong winds are only part of the story, and perhaps the most damaging aspect of hurricanes is the potential for flooding, partly from heavy rain and partly from the so-called *storm surge*, which is a dramatic rise in sea level that accompanies the fall in surface pressure in the centre of the storm. The surge is quite a simple process to understand. At the centre of the storm, the barometric pressure can be very low, perhaps lower than 900 millibars in some extreme cases. With less air pressure pushing down on the ocean surface, the sea level rises by about one centimetre (0.4 inches) for every millibar drop in pressure. So a rise in sea level of about one metre (40 inches) is not unusual for a big storm. Add to that the effect of large wind-driven waves and its clear that low-lying coastal regions are at great risk.

Hurricanes and typhoons are classified on the *Saffir-Simpson* scale of intensity, with five levels of intensity based on sustained wind speeds.

Hurricanes and typhoons are capable of producing *prodigious* amounts of rain. Half a metre (20 inches) of rain in a 24-hour period would not be unusual for a moderately strong storm, so it is no wonder that widespread flooding almost always

Storm category	Wind speed (mph)	Typical damage
1	119–153 km/h (74–95 mph)	Damage to mobile homes, trees and signs. Some minor coastal flooding.
2	154–177 km/h (96–110 mph)	Roofs damaged and some trees uprooted. Small boats at anchor break moorings.
3	178–209 km/h (111–130 mph)	Structural damage to poorly constructed buildings. Large trees blown down. Coastal flooding destroys small structures.
4	210–249 km/h (131–155 mph)	Many roofs blown off houses, mobile homes destroyed. Major flood damage to ground floors of all coastal buildings.
5	Greater than 249 km/h (>155 mph)	Complete destruction of roofs on many buildings. Extensive damage to windows and doors. Flood damage to all buildings within 460 metres (1,500 feet) of the shoreline.

Table 6.1 The Saffir-Simpson scale of hurricane intensity.

accompanies a land-falling hurricane. On some occasions the rain can be even heavier. In 1975, Typhoon Nina sat over the coast of China for three days and dumped over one and a half metres (60 inches) of rain, producing flooding that led to an astonishing 100,000 deaths when a dam burst. The location of landfall can have a big impact on the degree to which the flooding leads to disruption and loss of life. When hurricane Mitch hit Honduras in 1998 it downgraded in strength quite rapidly as it passed inland, but still produced exceptionally heavy rain in a region where many people lived in rather flimsy houses built on steep hillsides. The resultant flooding and landslides led to around 10,000 deaths.

Insight – Hurricane Katrina

Hurricane Katrina in August 2005 inflicted an almost direct hit on New Orleans, a city built largely below sea level. Although the winds were pretty strong (Katrina was a category 3 storm when it made landfall although it had reached category 5 while still over the sea), by far the biggest problems were caused by the floods. The banks built to hold back the waters of the sea and the Mississippi river broke in several places, flooding extensive areas of the city – largely the poorest parts. Drowning caused some deaths, compounded by problems of dehydration and heat stress in the days that followed the storm among the people who had not been evacuated. In all about 1,700 people died. This number was very high for a land-falling hurricane on the mainland US. Over recent years, improved forecasts and evacuation procedures have greatly reduced the risk to human life from hurricanes in America, but of course it is not possible to protect against damage to buildings, infrastructure and crops, and the cost of a land-falling hurricane in the US usually runs into billions of dollars.

The USA, Caribbean and the West Pacific are not the only places to suffer from damage from hurricanes, typhoons and tropical storms. One place that is particularly susceptible is Bangladesh, a largely low-lying country in South Asia formed from the delta of the river Ganges, at the head of the Bay of Bengal. It has a population of 125 million people squeezed into an area about the size of England, many of whom live only a few metres above sea level. In the Indian Ocean, tropical storms are known as cyclones, and even quite weak storms can lead to widespread flooding and loss of life. The Bay of Bengal is a formation region for *monsoon depressions*, the small but powerful storms that bring much of the monsoon rains to northern India, Bangladesh and Burma. To make matters worse, the shape of the Bay of Bengal, narrowing from south to north, acts like a funnel for the storm surge, forcing sea level to rise even further as the surge tracks northwards. In late April/early May 2008, a strong cyclone called Nargis developed in the Bay of Bengal and swept across the low-lying region of Burma at the mouth of the Irrawaddy river. Flooding was widespread across this area, and although there are no official statistics on the number of people affected by the storm, it is thought that well over 100,000 people lost their lives, with many more people being displaced from their homes.

There are two major forecasting centres that deal with tropical storms. Atlantic, Caribbean and East Pacific storms are covered by the National Hurricane Center in Miami, Florida, and storms in the West Pacific and Indian Oceans are dealt with by the Joint Typhoon Warning Center, located at the US Navy base at Pearl Harbor, Hawaii. Both centres issue advisories on individual storms every three hours, with updates on significant developments such as storms making landfall, or changes in storm strength. Since the advent of high-quality satellite images, our ability to forecast hurricanes has improved immensely. Prior to the satellite era, forecasters relied on reports from shipping, islands and aircraft, particularly the US Navy's 'hurricane hunters' who fly missions to investigate at close quarters any developing storm. It is true to say nowadays that a hurricane or typhoon will never catch us completely by surprise. The telltale signs in the satellite pictures are usually clear several days before a storm reaches hurricane strength. What is more tricky is predicting the precise track that a storm will take, and where it will make landfall. Some storms follow fairly straight tracks, tracking from east to west through the trade-wind belts. Other storms take very convoluted paths, changing direction completely in a short space of time. Another problem for the forecasters is that of predicting changes in strength of the storms. Accurate measurements of surface pressure and wind speeds *within* the storms are pretty sparse. Once again, forecasters rely on careful analysis of cloud patterns on satellite pictures, particularly around the centre of the storm and in the eye. Computer forecasting models are also getting more skilful at predicting storm tracks and intensities, but for the people living under the predicted path of major storms, there's always an element of 'will it hit or won't it?' and the clear, calm conditions which often precede hurricanes can create a false sense of security.

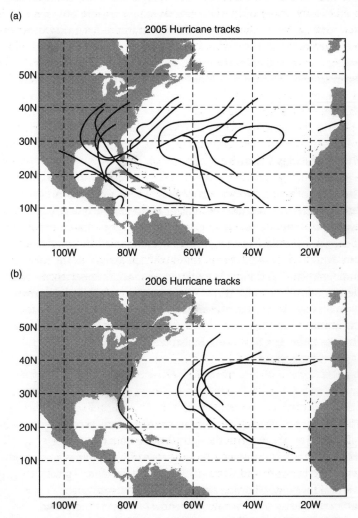

Figure 6.2 Hurricane tracks for the Atlantic hurricane season in (a) 2005 and (b) 2006. 2005 was a very active season, whereas 2006 was much quieter. Only storms which reached hurricane strength are shown, but they are shown for their entire track, even when below hurricane strength.

There can be large variations in the number and intensity of storms in any given season. 2005 was an exceptionally active season in the Atlantic. Of 31 named storms, 15 made it to hurricane strength and three of those went on to reach category 5. This was followed in 2006 by a rather quiet season in which there were only five hurricanes, most of which stayed well out in the Atlantic Ocean and didn't make landfall anywhere (see Figure 6.2). Several forecasting groups try to predict in advance how the forthcoming Atlantic storm season will turn out, looking at some of the factors detailed in Chapter 4.

Thunderstorms and hail

Thunderstorms are hardly rare events. There are few places in the world where thunder does not occur on at least a few days per year. Many thunderstorms amount to little more than a few claps of thunder and flashes of lightning, accompanied by some heavy rain. Much of the wet season rain in Africa arrives in the form of thunderstorms, and Florida experiences more thunderstorms than the tornado-prone mid-West. What makes some thunderstorms more intense than a short-lived storm that may be dramatic for a few minutes but then soon dies away? And why do some storms produce large and damaging hailstones?

What causes thunder?

Thunderstorms develop as a consequence of rapidly ascending plumes of air which carry heat and moisture from the surface high up into the atmosphere. The presence of such plumes is indicated by the development of *cumulonimbus* clouds – great vertical towers of billowing clouds several kilometres deep, often capped by wispy 'anvils'. Somehow (and as yet there is no consensus as to how), these rapid updraughts also result in a separation of electric charge within the cloud, with positive charge being concentrated near the top of the cloud, and negative charge building up in the lower parts. The ground beneath the cloud develops a positive charge, and if the charge difference between the cloud and the ground becomes large enough (of the order of one million volts per metre), the natural insulation of the air will break down leading to an electrical discharge between cloud and ground, visible as a bolt of lightning. Thunder is simply the result of the air through which the lightning bolt passes heating up to many thousands of degrees and expanding rapidly, creating a shockwave that propagates away from the lightning bolt. This makes it possible to determine how far away the lightning is, since the shockwave (i.e. the thunderclap) travels at the speed of sound (330 metres per second) whereas we see the lightning pretty well instantaneously however far away it is. So if the thunder is heard five seconds after the lightning is seen, the storm must be about 1650 metres (approximately one mile) away.

The main factor that distinguishes severe thunderstorms from the more typical, brief summer afternoon storm is lifetime. Regular thunderstorms are over quite quickly, and to some extent actually bring about their own downfall. A thunderstorm develops in response to strong vertical convective air currents called *updraughts*, bringing about condensation and forming cloud. If the cloud extends high enough in the atmosphere then it will contain ice crystals, leading to the typically wispy appearance of the 'anvil' at the top. These ice crystals can grow rapidly and form precipitation. Often they melt on the way down and fall to the ground as rain. Once precipitation begins, intense *downdraughts* of air start to form, bringing cold air down from high above the ground. If these downdraughts coincide with the inflow of warm air at the base of the cloud that forms the updraughts then they can actually cut off the heat and moisture supply to the storm and kill it off, as shown in Figure 6.3. This type of storm is often referred to as a *single cell storm* as it consists of a single convective cell, with one main updraught and downdraught. Colour plate 14 shows the typical appearance of such a storm.

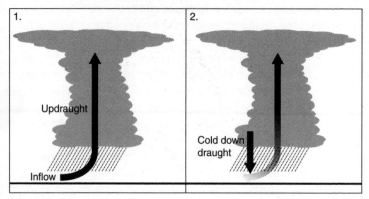

Figure 6.3 *Life cycle of a single-cell thunderstorm, showing how the cold downdraught can cut off the flow of air into the base of the storm, thus killing off the storm.*

In really severe thunderstorms, this self-destruction mechanism doesn't operate, allowing the storm to continue to develop and even regenerate. If the wind speed and direction changes sufficiently with height in the vicinity of the storm, the storm will be tilted and twisted in the vertical, meaning that the downdraughts do not interfere with the inflow of air at the base of the storm which feeds the updraught. Plate 15 shows this type of storm developing and the vertical tilt with height can be clearly seen. New convective cells can also be seen developing on the near side of the storm. In some cases, the cold air in the downdraught will spread out on hitting the ground, causing an enhancement of the surface wind known as a *gust front*. This gust front often runs just ahead of an advancing storm, explaining why the wind often increases suddenly a few minutes before the heavy rain or hail arrives. The cold, dense air in this gust front can actually act to force the warmer air at the surface to rise, leading to the formation of another convective updraught adjacent to the original cloud – a process shown in Figure 6.4. In the right conditions, this mechanism can repeat itself many times, so the storm can become self-sustaining and quite long lived. This type of storm can generate large hailstones.

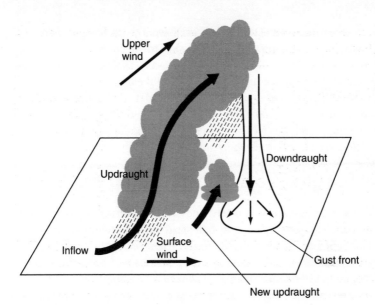

Figure 6.4 Multi-cell thunderstorm. The vertical tilt of the storm due to vertical windshear leads to the downdraught being separated from the inflow, thus prolonging the life of the storm. As the downdraught hits the surface, it spreads out as a gust front, lifting warmer air to form a new updraught.

What causes hail?

As with the causes of thunder, the exact mechanisms for the formation of hail are still an area of active research. Two factors though are fairly clear. The first is that a considerable portion of the storm cloud must be at temperatures of below freezing so that ice particles are able to form. The second factor is that for large hailstones to form, they must remain suspended within the storm cloud for a considerable period of time rather than falling out under the action of gravity while still very small. This second factor requires very strong updraughts of air which can counteract gravity, holding initially small ice particles stationary or sweeping them up through the cloud where they can grow by the accumulation of further ice or liquid water which then freezes on contact with the growing hailstone. Some hailstones are believed
(Contd)

to have made several transits up and down a cloud, being carried by an updraught, growing until they start to fall under the action of gravity and then being swept back upwards again if they fall into another vigorous updraught. Some of the biggest hailstones observed are formed by several smaller stones sticking together, possibly as their surface layers start to melt in the warmer parts of the cloud, and then refreezing as they are swept up again into the colder regions of the cloud by another updraught.

The size of hailstones is usually described in terms of either sporting equipment, e.g. golf-ball sized, or food items, e.g. pea-sized. This type of hail can be quite nasty if you're caught out in it, but grapefruit-sized hail is getting pretty serious and potentially life-threatening. Hailstorms can cause serious damage to buildings, vehicles, aircraft and of course people. But perhaps one of the most devastating impacts of hail is on crops. Hail can very effectively strip the ears from a growing cereal crop, or knock growing fruit off its trees. A large number of smaller hailstones destroy crops more effectively than a few large hailstones. In 1788, a long-lived hailstorm swept across a large swathe of agricultural land in northern France, destroying a considerable portion of the wheat crop just prior to the harvest. Wheat prices rose as a consequence, making flour and therefore bread more expensive. Unrest over bread prices in Paris was a factor in the initiation of the French Revolution, so it could be said that the Revolution and subsequent Republic was triggered in part by a hailstorm.

Forecasting the occurrence of severe thunderstorms is still a developing area. Computer forecasting models are now quite good at predicting the large-scale conditions that are likely to generate heavy showers and thunderstorms, even several days in advance. What is much less predictable is the exact location, timing and duration of any storms. Forecasters rely on rainfall radar to pinpoint the position of showers as they develop, but it is still difficult to say which particular showers will die away quickly and which will develop into something more severe. Careful monitoring of the situation as it develops allows forecasters to assess what will happen in the next few hours – a process known as *nowcasting* (see Chapter 8 for more details), but beyond that it's not possible to do more than issue warnings of the potential heavy showers over a region.

Floods

Flooding comes in several different forms. We've already looked at the flooding associated with tropical cyclones, including the effect of storm surges. This type of flooding can be rapid, quite widespread in low-lying coastal areas, and have devastating effects. On a much longer timescale, widespread flooding can occur as the result of extended periods, usually many weeks, of wetter-than-average weather. As groundwater levels rise, land gradually becomes saturated and rivers fill their channels to the top of the banks. At this point one or two particularly wet weather systems can push things over the edge leading to extensive flooding along the flood plains of river systems. In places where settlements have been built on the flood plains, this can result in extensive damage to property, but often much of the damage is to agricultural land and crops, and because this type of flooding is more common during the winter season when there are fewer standing crops in the fields, the damage isn't excessive. In fact some agricultural areas have traditionally relied on seasonal flooding to spread fertile sediment on the fields to maintain their productivity.

Much more dramatic is sudden and unexpected flash flooding brought about by single extremely heavy rain events. This is perhaps more of a summer phenomenon, with convective storms of the type described earlier being largely responsible for the heavy rainfall amounts. Local geography is also a major factor in flash flooding. If an exceptionally heavy thunderstorm occurs over a fairly flat region, perhaps in the lower catchment of a major river basin, the water will be spread fairly evenly over the surface and will make its way fairly slowly into the river. If the soil is already saturated then there may well be standing water in the fields, but there will be no dangerous and damaging torrents of water if there are no steep slopes or narrow valleys to focus the flow of water into the main river system.

If a concentrated, heavy thunderstorm occurs over a small, steep-sided river catchment, the results can be devastating. Water that is unable to soak into the soil will run quickly down the steep sides of the valleys, channelling quickly and efficiently into the narrow streams that feed the main river. A swollen river flowing down a moderate gradient can have tremendous force – enough to move large boulders and uproot trees near the bank. In mountainous regions this type of flood can be excessively dangerous to walkers, climbers, canoeists, etc. in the river valley, particularly if the storm has occurred further up the river and the people are actually unaware that a flash flood is on the way. This type of flooding really hits the headlines when it occurs in a populated valley and involves damage to property on an awesome scale. Once again, mountainous regions are most at risk, particularly in the tropics, where rainfall amounts can run into metres per day. Steep-sided islands like the Philippines and Indonesia suffer fairly regularly from damage to property, infrastructure and loss of life by flash flooding, and the mountainous states of central America are also familiar with this type of disaster. The mid-latitudes are not immune to these events either. Mountain villages in the Alps and other mountain ranges in Europe are all at risk from flash flooding, and even relatively low-lying southern Britain has experienced flash floods.

Insight – The 1952 Lynmouth flood

One of the most devastating examples of flash flooding in the UK occurred on the 15/16 of August 1952, when many buildings in the village of Lynmouth on the north Devon coast were destroyed and 34 people were killed by flash flooding of the East and West Lyn rivers. This was caused by fairly widespread heavy rain across the south-west of England, exacerbated by a wet summer meaning that the ground was already fairly saturated. Although the heavy rain was widespread, it was focused into a flash flood by the steep-sided river valleys that feed the main river running through Lynmouth which stands in a small gap in the cliffs along the coast. Much of the rainfall in a catchment a few tens of kilometres square was funnelled through that gap. The loss of life was probably made worse by the fact that the flooding occurred overnight. Many people didn't realize what was happening until the flooding was already very severe, and were then trapped in their collapsing homes.

A similar event to the Lynmouth Flood occurred in 2004, coincidentally also on the 16 of August, 50 miles down the coast at Boscastle in Cornwall. In this case, the heavy rain was not widespread, but instead occurred in a narrow line of intense showers just inland from the coast. The showers themselves were triggered by convergence of winds flowing around a headland to the west of Boscastle. Photographs and video shot by holidaymakers on the beaches of north Cornwall that day show clears skies out to sea and a great dark wall of thunderclouds on the inland side. While individual showers moved quickly along this line, new ones were generating continually at the western end, leading to virtually continuous heavy rain along a very narrow band over a seven-hour period. The catchment of the river Valency, which meets the sea at Boscastle, lay directly beneath this band, and once again the water was funnelled quickly into the river by the steep-sided valleys. Parked cars added to the natural debris washed down by the river and many buildings were damaged or destroyed completely by the flood. Dozens of people had to be rescued by helicopter from the roofs of buildings completely cut off by the flood water. On this occasion no lives were lost but the devastation of a flash flood, captured live on television, was an awesome spectacle.

Tornadoes

When it comes to extreme weather, tornadoes are perhaps as extreme as it gets. All tornadoes form in association with intense, organized convective storms, often involving thunder and lightning, hail and towering black clouds. Nothing gets weather enthusiasts as excited as tornadoes, probably due to a number of factors. The first is unpredictability; while the parent thunderstorms themselves may be quite easy to forecast, it is almost impossible to predict more than a few minutes in advance which storms will spawn tornadoes and where. A second factor is scale; if you happen to be 'lucky' enough to be in exactly the right place at the right time, it's possible to watch a tornado destroying trees and buildings perhaps a few hundred metres away, and be completely safe, but always with the risk (some would call it a thrill) that the storm will change direction and head towards you. The third factor is that of power. Tornadoes pack the strongest winds on Earth, with speeds that are almost impossible to measure since a powerful tornado will destroy any standard measuring device it encounters. The strength of tornadoes in the US is measured on the Enhanced Fujita-Pearson (EF) scale. This scale is based on surveying the damage caused by the tornado due to the difficulties of getting any kind of in-situ measurements of wind speed. The strongest category, EF5, can destroy brick buildings and carry cars through the air for distances of over 100 metres. Winds in a tornado of this category are assumed to be gusting at speeds in excess of 320 kilometres per hour (200 miles an hour).

Regular occurrences of damaging tornadoes are almost exclusively limited to the continental USA but that's not to say that they don't occasionally occur elsewhere. Sightings of funnel clouds are fairly common in the UK, and occasionally they can develop into full-blown tornadoes. In July 2005 a tornado passed over parts of Birmingham in the English Midlands causing extensive but localized damage to buildings, trees and cars. Other places such as northern India are also susceptible to damaging tornadoes.

However, it is over the US, and particularly the states of the Great Plains, where damaging tornadoes are a regular occurrence. Florida may get more thunderstorms per year, but the combination of hot, dry air flowing down from the Rockies and high deserts to the west, and warm, moist air flowing up from the Gulf of Mexico to the south means that the area known as 'Tornado Alley' (see Figure 6.5) sees the ideal conditions for tornado formation more often than anywhere else on Earth.

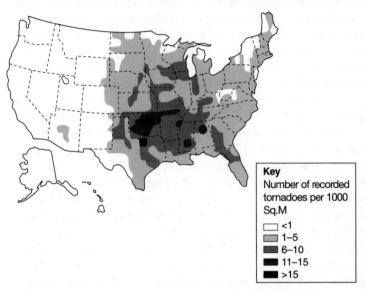

Figure 6.5 Map showing the typical frequency of occurrence of tornadoes in the USA.

The peak season for tornado formation in this part of the world is late spring to early summer. In the right circumstances, the hot dry air from the west flows *over the top* of the warm, moist air flowing up from the Gulf of Mexico. Daytime heating of the surface makes the air mass unstable at the surface, but deep convection is initially prevented by the capping layer of hot, dry air that acts as a lid to rising thermal currents of air. This situation can persist for several hours as energy builds up within the atmosphere but cannot be released due to the so-called *capping inversion*. Rising currents of warm, moist air gradually eat away at this inversion, and eventually a thermal may break through, leading to the rapid development of an intense storm within a few tens of minutes. This by itself is not enough to guarantee the formation of tornadoes, but it provides the right conditions in which they may form. The important ingredient to actually produce tornadoes is rotation. As we have already seen, the rotation in larger-scale weather systems such as hurricanes and mid-latitude depressions is provided by the Coriolis effect, as air which is initially moving towards the centre of a weather system is deflected from its path by the rotation of the Earth. This is not the reason why rotation starts to develop within a thunderstorm, as the size of the storm is too small to be much affected by the Earth's rotation. The factors that lead to storm rotation, and eventually to tornadoes, are still subjects for intensive research, much of it carried out at the National Severe Storms Laboratory, right in the heart of tornado country at Norman, Oklahoma. Much of the work is focused on the use of *doppler radars*, which can detect the direction of motion of air within a storm as well as the rainfall intensity. These instruments can now be made small enough to be mounted on a vehicle. They can be driven right into the heart of thunderstorms, accompanied by other vehicle-mounted equipment and human observers to get as close a view as possible of developing tornadoes. One possible mechanism for the generation of rotation in storms is that, in and around a thunderstorm, there may be several horizontal currents of air blowing in very different directions in close proximity. These opposing streams of air may generate rotation about a horizontal axis which is then tilted into the vertical by strong convective updraughts. Clearly very detailed measurements of the air flowing around a storm are needed to be able to confirm or deny such a theory.

Given the unpredictable nature of tornadoes, and the fact that we don't fully understand their formation, what chance do we have of being able to forecast them with sufficient lead time to prevent loss of life? It is certainly possible to forecast, at least a day ahead, the occurrence of the large-scale conditions in which tornadoes *may form*, and this information can be imparted through the media as a *tornado watch*, which will often cover several states in the mid-West – a very large area. As the situation evolves, forecasters can monitor and pinpoint the development of individual storms using satellite and radar imagery, together with reports from observers on the ground. Specific warnings of tornado formation for individual counties can now be issued 20 minutes in advance, which is enough lead-time for most people to be able to find shelter or avoid the area. Of course, no forecast will be able to prevent extensive damage to buildings, trees, infrastructure and crops by a strong tornado. The area most at risk from tornadoes is not densely populated and many storms miss human habitation altogether. However, there will always be occasions when severe storms hit major centres of population.

Insight – The 1999 Oklahoma City tornado outbreak

One of the most devastating examples of tornadoes affecting a built-up area occurred on 3 May 1999. Dozens of tornadoes were spawned across the state of Oklahoma, not in itself an unusual occurrence since Oklahoma sees more tornadoes per year than anywhere else in the US. However, on this occasion several densely populated suburbs of Oklahoma City were hit by tornadoes ranging up to EF5 in strength. In the suburb of Moore, over 1,000 homes were *completely* destroyed by one particularly strong and long-lived tornado and over 4,000 more suffered damage. The repair bill ran to over one billion US dollars and 40 people were killed in and around Oklahoma City.

Snow and ice

Does snow count as severe weather? It depends entirely upon location. In some parts of the globe, snowfalls of several metres during a winter season are completely normal and wouldn't be regarded as extreme at all. In other places, where perhaps a fall of snow isn't even an annual event, even a few centimetres can lead to normal life grinding to a halt. Take the example of the 30 January 2003 in south-east England. The forecast of this case is discussed in more detail in Chapter 8, but with 12 hours warning, the location, timing and amounts of snow were reasonably well predicted. Accumulations of snow were no more than a few inches in most places. Unfortunately they coincided with the evening rush hour and much traffic was leaving London and heading north and east, into the area of heaviest snowfall. This being south-east

England in the twentieth-first century, whole winters can pass with no significant falls of snow at all, so people were largely unprepared for this type of severe weather, and many chose to ignore, or did not hear, the forecast. In continental Europe, where snow during winter is common, people habitually fit snow tyres to their cars in winter. They know how to modify their driving in snowy and icy conditions. Highway authorities invest large sums of money in snow clearance and ice prevention equipment and are on standby to use it. In the south-east of England this is not the case. Numerous minor accidents on the untreated icy roads that evening led to paralysis of the road network to the north-east of London. With all the lanes of the major roads blocked by stationary traffic it was impossible for recovery vehicles and snow-clearing equipment to get to the affected areas. Thousands of people were stranded overnight in freezing conditions, many without any warm clothing, food or blankets. A similar weather event in continental Europe, Scandinavia, the northern US or Canada would have passed without incident.

What causes snow?

There is nothing particularly special about the conditions needed for snow to fall. Outside the tropics, much of the rain that falls at the Earth's surface started off within the clouds as ice particles. The key to whether these frozen particles fall to the ground as snow or melt on the way down leading to rain is the temperature of the lowest part of the atmosphere. Clearly if the temperature of the lowest few kilometres of the atmosphere is well above freezing then snow falling through this layer will turn to rain well before it reaches the ground. However, if this lowest section of the atmosphere is all below freezing then any precipitation will fall as snow. Forecasting difficulties arise when the lowest part of the atmosphere is just above freezing, giving the possibility that the snow may or may not be quite melted before it reaches the ground.

In cold countries where snow is common, snow can still count as severe weather but there has to be a whole lot more than a few centimetres before it starts to cause problems. The northern, central and north-eastern states of the US are certainly used to heavy snowfalls during the winter season, but the position of the jet stream and the influence of the Rocky Mountains mean that freezing cold and mild conditions can fluctuate throughout the winter. One problem which particularly affects the north-eastern states is that weather systems that start out by bringing rain can turn rapidly to snow as cold air from the continental interior is dragged into the system. Another characteristic of snowy weather in the US north-east are so-called *lake effect* snow events. These occur when cold northerly or north-westerly winds blow over the unfrozen waters of the Great Lakes, picking up moisture which then falls as snow on the lee shore. Cities such as Cleveland, Ohio on the southern shore of Lake Erie and Rochester, New York to the south of Lake Ontario are familiar with this effect, and snow amounts in these areas can be very substantial.

Insight – The Valentine's Day storm

One case that exemplifies the often rapid transition from warm to cold weather in the US north-east, and also the obsession in the media for naming weather events, was the Valentine's Day storm of February 2007. Up until early February, the winter had been largely mild across the north-east. However, on 10 February a depression formed in the lee of the Rockies, which then tracked eastwards and merged with another depression that had formed off the eastern seaboard. The combination of the cold air being pulled down from the north behind the Rocky Mountain system, and the warmer moist air associated with the coastal low led to extensive snowfalls from Iowa across to New York and the Canadian Maritime Provinces. Lake effect snow added to the amounts falling to the south-east of Lakes Erie and Ontario, with Rochester NY receiving 53 cm (21 inches). The heaviest snowfall associated with the storm was the 122 cm (48 inches) that fell in the Vermont ski resort of Stowe. Further south,

(Contd)

the storm spawned an unseasonable tornado outbreak across the Gulf States. Sadly, 37 people died as a result of the storm, most of them in road traffic accidents. Many major roads were closed by accidents or heavy snow, and many people were stranded in cars for over 24 hours. Several major airports were forced to close due to the conditions, and at JFK airport in New York, some planes were actually frozen to the tarmac. Many schools and colleges closed down, and rather ironically, a government committee meeting on global warming had to be cancelled.

One weather type which is also associated with cold weather, particularly in the northern and central US, is the *ice storm*. In the UK ice storms are known more prosaically as *freezing rain* events. Ice storms occur when rain falls onto surfaces which are colder than 0°C (32 F), and freezes on impact. The classic conditions for an ice storm are a warm front passing over an area that has previously experienced several days of cold, clear weather resulting in surface temperatures which are below freezing. Most precipitation during winter starts off in the cloud as ice, but can melt if it falls through a layer of warm air on its way down. If it is warmed to just above freezing, but then lands on a surface which is below freezing then it will freeze again almost instantaneously, resulting in the build-up of a layer of clear ice on the surface. In extreme cases this accumulation can be several centimetres thick, and can bring down power lines and large branches from trees. Road surfaces and pavements become treacherous ice rinks. Perhaps the most dangerous circumstances occur when this layer of clear ice is only a few millimetres thick. Because it is clear it is almost invisible, and in this case it is known as *black ice*. Drivers can be fooled into thinking that the roads are free of ice resulting in many minor and occasionally much more serious accidents. The Valentine's Day storm of 2007 led to extensive ice accumulations along the US Atlantic States, with coastal New York, Maryland and Virginia being particularly badly affected.

Forecasting snow and ice can be very difficult depending on circumstances. In some cases it can be very obvious that snow is on the way, particularly if temperatures are clearly well below freezing. Things are much more difficult if the temperature is hovering just above freezing. In such cases precipitation that starts off as rain can soon turn to snow. This transition is brought about by the rain itself. As rain falls from a cloud into air that is not completely saturated, the raindrops start to evaporate. This process cools the air, and can bring the air temperature down below freezing. At this point the raindrops themselves can start to freeze, forming snow or ice pellets. Many widespread snow events occur in association with a warm front that comes up against a mass of cold air over a continental land mass. The progress of the front can stall as it runs up against the wedge of cold, dense air, resulting in snow that may originally have been forecast to turn to rain as the warm front moves in. However, with the front becoming stationary, the transition never occurs and the snow persists. Situations like this are sometimes not well handled by computer models, and forecasters often rely on information from radar and radiosonde balloons which can provide information about the nature of the falling precipitation and the temperature structure of the atmosphere.

Insight – Is it ever 'too cold to snow'?

In particularly cold weather conditions, a phrase you may hear is 'It's too cold to snow'. This seems an odd statement, for surely cold weather is precisely what is needed for snow to fall. The exact origins of this phrase are not entirely clear, but it may originate from regions where the coldest weather occurs during wintertime anticyclonic conditions. During winter, a persistent anticyclone can lead to periods of many days of clear skies and light winds allowing the surface of the Earth to cool strongly by radiation of energy to space. Night-time temperatures can plummet to many degrees below freezing, and the very weak winter sunshine does little to warm things back up during the day. In such conditions any sort of precipitation is unlikely, and the arrival of any sort of precipitation-bearing weather system will probably be preceded by increasing cloudiness, winds picking up and a general rise in temperature.

Drought

Drought isn't so much severe weather as severe lack of weather. For many of us, it is only when it stops raining for a significant length of time that we realize how crucial rain is to our existence. It is hard to come up with a single definition of drought as it varies so much with location.

Drought hits hardest in places where the population depends upon small-scale, localized farming to produce food to sustain the population. Perhaps the starkest examples of this in recent times have been the extended drought periods, sometimes lasting many years, in sub-Saharan Africa. This part of the world depends on wet season rains brought by the northward excursion of the ITCZ during the summer months. The ITCZ over Africa is not a continuous band of rain, but an agglomeration of small-scale weather systems and convective showers that produce localized downpours. During droughts, these systems are less frequent, smaller in extent or non-existent. There's a cruel positive feedback that worsens drought in such semi-arid areas too. Much of the moisture that supplies any given rainfall event here comes from evaporation from the soil of water that fell in previous events. Once the soil dries right out, this source is removed and so even if the large-scale conditions of the atmosphere are right for producing weather systems, there's no water for them to feed off.

One place where a massive population is susceptible to drought is India. Until fairly recently, India had an overwhelmingly agricultural economy with subsistence farming supporting a sizeable proportion of the population. Even today, several hundred million people still depend on farming. This agricultural activity depends largely on the abundant rain of the monsoon season (June to September). There are some irrigated crops grown during the drier seasons, but the stores of water that feed them are filled during the monsoon, so the productivity still relies on the wet season rains. The Indian summer monsoon is a remarkably stable system. The rains never fail completely over the country as a whole and departures from the seasonal normal rainfall larger than 20 per cent are extremely rare. Since 1870 there have only been five years in which the monsoon rainfall total has fallen below 20 per cent of normal for the country as a whole. In a typical drought year the centre of the country tends to be very dry, with the main rain band retreating into the hills and mountains of the north and the very south of the peninsula.

Of the five most extreme drought years since 1870, four were also years in which an El Niño was occurring in the tropical Pacific Ocean, and the onset of El Niño has historically been used as an indicator of a potential monsoon drought. However, in 2002 the monsoon rains were deficient by more than 20 per cent in a year with no El Niño. The main deficit occurred during July, when the monsoon went into an extended *break period*, catching the forecasters by surprise. July is usually the month of peak rainfall during the monsoon and the rain at this time is crucial for the development of many different crops. The failure of the meteorological agencies to forecast this drought prompted a complete overhaul of the forecasting methods used to predict the strength of the monsoon by the Indian meteorological department. It was clear that the influence of El Niño had been diminishing over the previous 20 years or so and was no longer such a reliable precursor. Another example of this change had occurred in 1998, when one of the strongest El Niño events on record failed to influence the strength of the monsoon, with the rainfall during this season being slightly above normal. Whether this shift in influence is due to climate change or natural variations in the patterns of climate connections is still a question for further research. Whatever the reason, it means that one of the more reliable indicators of the strength of the monsoon is not at all reliable any more.

El Niño is known to affect rainfall amounts in places other than India. One place where the rains are also strongly in thrall to El Niño is eastern Australia, which is home to much of the country's population. Droughts in this region not only have a detrimental impact on agriculture and domestic water supply, they can also lead to widespread wildfires, which near populated areas can also endanger life. The summers of 1982–3 and 1997–8 (both

El Niño years) saw drought and widespread fires near cities such as Canberra and Melbourne. However, since 2003, the climate of south-eastern Australia seems to have settled into a near-permanent drought state which seems unconnected with events in the tropical Pacific. Agriculture in the Murray-Darling river basin has suffered badly. Again there is speculation that climate change is taking a hand and upsetting the normal climate patterns. This is a particularly pointed debate since the Australian government over the past ten years or so has taken a rather sceptical view of climate change, and has refused to sign up to the Kyoto protocol (see Chapter 10). In April 2006, following a summer during which there had been a weak El Niño in the Pacific, climate scientists declared that the extended drought had reached the scale of a once in 1,000 years event. Prime Minister John Howard even suggested that the population should start praying for rain. On a more practical level, several states and cities have started to review how they use water resources, how they store water and whether current industrial and agricultural practices are sustainable. During 2007, the El Niño conditions in the Pacific switched over to La Niña, which has historically led to greater than normal rainfall in Australia. However, in September 2007 the Australian Bureau of Meteorology issued a stark statement. Rainfall amounts had not recovered – the first time since records began in 1900 that an El Niño-induced drought in the Murray-Darling basin had not been followed by at least one three-month period of above-average rain. The bureau suggested that several years of above-average rain would be necessary to return water resources to normal. While it would be very difficult to prove that the current drought was due to climate change, it is certainly the case that these conditions are outside the variability seen in the previous 100 years.

Summary

Extreme weather, wild weather, freak weather. Call it what you like, it is this type of rarely occurring event that hits the headlines, affects people's lives and sometimes leads to widespread loss of life. These types of event are more difficult to forecast by the very fact of their rarity, but they are also the events where an accurate forecast can really make a difference between life and death.

A good weather forecast must always start from good observations of the weather, and in the next chapter we'll start to look in detail at how the weather is observed and recorded. Then in Chapter 8 we'll spend a lot more time looking at how the big weather forecasting agencies go about producing their forecasts.

10 THINGS TO REMEMBER

1 *The very definition of extreme weather varies from place to place. Conditions that might be considered extreme in one place may well be normal somewhere else.*

2 *The development of intense mid-latitude depressions can often be spotted in satellite pictures by the presence of an 'emerging cloud head'.*

3 *Tropical storms such as hurricanes form exclusively over warm tropical oceans.*

4 *Although tropical storms produce very strong winds, much of the damage they cause when they make landfall is due to flooding from heavy rainfall and the storm surge caused by the sea surface bulging upwards as the atmospheric pressure falls in the centre of the storm.*

5 *Thunderstorms consist of intense updraughts of air that can carry cloud droplets and ice particles high into the atmosphere. These updraughts lead to the separation of electric charge within the cloud. A lightning strike (and the accompanying thunderclap) occurs when the charge difference becomes so large that a massive electric current flows through the cloud.*

6 *Tornadoes are perhaps the most extreme weather systems on Earth. Although it's hard to measure the winds in a tornado, the damage done by a really big one suggests that the winds may be gusting up to 200 miles per hour.*

7 *A unique combination of airflows from different directions, with hot dry air from the High Plains flowing over warm moist air from the Gulf of Mexico means that regular, large tornadoes occur almost exclusively in the central plains of the US during spring and early summer.*

8 *Most of the precipitation that falls outside of the tropical regions of the Earth starts off within the clouds as snow. Whether it is still frozen on reaching the ground depends on the temperature of the lowest part of the atmosphere.*

9 *An 'ice storm' occurs when rain falls onto very cold surfaces and freezes on contact. Trees, power lines cars and road surfaces all become coated with a layer of clear ice which can be several centimetres thick, causing considerable damage and making driving conditions treacherous.*

10 *The definition of drought varies from place to place. A severe drought in the Indian monsoon, such as that during the 2002 season, can be caused by just a few weeks of deficit rainfall. In southeast Australia, rainfall amounts have been below normal for about a decade, leading to an extended drought which has affected farming practices and has led to a series of devastating wildfires in populated areas.*

7

Observing the weather

In this chapter you will learn:
- *about the history of observing the weather*
- *about some of the key developments in weather observing instruments*
- *how the main weather forecasting agencies monitor the weather from the Earth's surface and from space*
- *how to set up your own observing station and contribute your own data to weather observing networks.*

Weather observations through history

Man has always kept an eye on the weather. Going right back to prehistoric times, life was much more sensitive to the elements than it is in today's modern world. The weather conditions on a particular day must have affected how the day was spent, and may even have determined whether there was food to eat or not. Life was very much dictated by the march of the seasons and the daily fluctuations of the weather. It is almost certain that early man started to develop the skill of looking at the sky and sensing the wind direction in order to get some idea of how the weather would develop during the day. It is probably no coincidence that human life evolved originally in tropical regions where seasonal variations in weather are small. Mankind gradually spread into temperate regions, where there are larger variations in weather, only when

146

he developed the ability to adapt to variations in weather, for instance by making and wearing clothes. As farming developed, human activity continued to be sensitive to the weather and there is no doubt that ancient civilizations began to keep records of the weather. The ancient Egyptians for instance were very dependent on the annual flooding of the Nile to fertilize their fields and feed their irrigation systems, and they kept written records of the strength of those annual floods. Greek and Roman historical literature contains plenty of references to how the weather affected military campaigns, and the ancient Greek philosophers even began speculating on what actually caused the weather. Norse sagas also contain information on the weather, and its impact on exploration by boat. The English Anglo-Saxon Chronicle makes reference to the weather when it affected harvests or battles. Artwork can also be used by meteorological historians to assess how the weather and climate might have varied in the past. Pictures of 'frost fairs' on the frozen river Thames in London during the seventeenth century for instance help to confirm that the winters were more severe during this period.

Two key seventeenth-century developments

All manner of historical documents make reference to the weather, but up until the seventeenth century they all had one thing in common – they were descriptive rather than quantitative. The weather could be compared to the previous year, or the previous day, but there was no way of saying precisely how hot, cold, windy or wet the weather actually was. Extreme events such as the storm that destroyed the Spanish Armada in 1588 could be described in terms of the damage they inflicted, but there are no records to tell us exactly how intense the storm was as there were no instruments to measure it. Two inventions in the seventeenth century changed forever the nature of historical weather records. The first was the mercury barometer, designed by **Evangelista Torricelli** in 1644. This instrument used a column of mercury, the weight of which was balanced by the pressure of the air in the atmosphere.

The higher the air pressure, the higher the column of mercury it could support. Typical air pressure at sea level can support a column about 760 millimetres (approximately 30 inches) high, and the unit of *inches of mercury* is still used on many household barometers despite the fact that most barometers contain no mercury whatsoever, and the standard meteorological unit for measuring atmospheric pressure is the millibar or hectopascal. The second key invention was that of the thermometer by **Ferdinand of Tuscany** in 1660, although it was refinements of his original design that led to the mercury-in-glass instrument with which we are familiar today. Also crucial for the measurement of temperature was the development of a *scale* on which the temperature could be recorded. In the early eighteenth century the German scientist **Daniel Gabriel Fahrenheit** defined his temperature scale, with fixed points of 0 degrees Fahrenheit (F) at the temperature of a mixture of water, ice and salt, and 96 degrees Fahrenheit corresponding to human body temperature. At around the same time in Sweden, **Anders Celsius** defined 0 and 100 degrees Celsius (C) as the freezing and boiling points of water respectively. The Celsius scale is the one now adopted as a meteorological standard, but in the US, Fahrenheit is still the preferred scale. In Britain too, TV weather presenters often lapse into Fahrenheit Figures, especially when describing hot weather since 'temperatures soaring into the nineties' in Fahrenheit always sounds more impressively hot than 'temperatures rising into the thirties' in Celsius. Despite the development of standard instruments and scales for temperature measurement, it would be many years before there was an agreed standard for actually taking temperature readings. The value of temperature that a thermometer records will depend strongly on its *exposure*. A thermometer placed in direct sunlight will read a much higher temperature than one in the shade, and one attached to the wall of a building may record higher temperatures overnight than one suspended well away from any large structures as a building itself may act as a source of heat. It was only in the latter part of the nineteenth century that the standard method for recording temperature became the use of a slatted wooden box or screen to shade thermometers from direct sunlight while still allowing air to flow past the instrument.

LUKE HOWARD NAMES THE CLOUDS

Making objective measurements of clouds is far more difficult than recording pressure or temperature. The somewhat ephemeral nature of clouds, and their infinite variety, makes them very difficult to classify. Perhaps the easiest aspect to measure is the total cloud amount in the sky. Meteorologists divide the sky into eighths or *oktas*, with eight oktas indicating total cloud cover, and zero oktas indicating a completely clear sky. However, even the total amount of cloud is difficult to measure objectively, especially when the cloud is patchy with many small breaks.

Describing the *form* of clouds is even more difficult than assessing the amount present in the sky. Every cloud or skyscape is unique, and changes from minute to minute. However, the form of clouds can tell an experienced meteorologist a lot about how those clouds formed, and hence what type of weather they might be expected to bring. There is much more on how to assess the form of clouds to make weather forecasts in Chapter 9. A concise naming scheme for clouds would allow meteorologists to describe the sky to someone in a remote location who could then build up a mental picture of how the sky looked. If this information could be gathered from many different locations across a country or even wider area, then it could be plotted on a chart and used to build up a big picture of the distribution of clouds of different types.

In 1802, an English industrial chemist called **Luke Howard** devised exactly such a scheme. There is no doubt that he was inspired by the growing tendency in the scientific community to name and classify everything they studied in the natural world, from plants and animals to rock types. Following the work of the Swedish scientist **Carl von Linne** (usually known as **Linneaus**), scientists were developing classification systems based on groups and subgroups, with pseudo-Latin names as Latin was still the universal language of the intellectual world at the time. Howard was clearly fascinated by the forms that clouds took, and filled many sketchbooks with detailed pictures of cloud formations.

In his essay on 'Modifications of Clouds' Howard named four basic categories of cloud, giving them Latin names to mirror the naming conventions being used in the plant and animal kingdoms. His four types were:

- **Cumulus** *(from the Latin for 'heap')* – *Convex or conical heaps increasing upward from a horizontal base.*
- **Stratus** *(from the Latin for 'layer')* – *Widely extended horizontal sheet.*
- **Nimbus** *(from the Latin for 'rain-cloud')* – *Systems of clouds from which rain falls.*
- **Cirrus** *(from the Latin for 'fibre')* – *Flexuous fibres extensible by increase in any or all directions. The fibrous appearance is due to these clouds being formed from ice crystals rather than water droplets.*

Acknowledging that classifying clouds was not as exact as classifying animal and plant types, Howard also defined intermediate categories in which a cloud could be a blend of two of the basic types, such as cirro-stratus or cumulo-stratus (nowadays we'd call it stratocumulus). Howard's naming convention for clouds gained widespread approval and is still the basis for the names of all the cloud types we use today. Further developments have led to more sub-categories. In particular the use of the prefix *alto-* has been introduced to refer to cloud in the middle part of the troposphere at altitudes of between about 2,000 and 7,000 metres (7,000 to 23,000 feet). This gives two extra cloud types; medium-level layer cloud is *altostratus*, and medium-level lumpy cloud is *altocumulus*. While altostratus tends to be associated with frontal systems and usually has a uniformly grey appearance through which the Sun may be weakly visible, altocumulus can take on many different forms, some associated with fronts (see plate 16) and some not (plate 17). Clouds above these heights will almost certainly be formed from ice crystals rather than water droplets and so will be of the cirrus type – *cirrostratus*, *cirrocumulus* or more often just plain *cirrus*. Plate 18 shows the typical appearance of cirrus clouds as fibrous tufts, often with a hook shape at one end. Low-level clouds have no prefix, so the three main types are *stratus, cumulus* and *stratocumulus* for a layer cloud with a lumpy appearance within the layer. The two main rain-bearing clouds are nimbostratus – the thick, dark grey layer cloud associated with weather fronts – and cumulonimbus, the towering convective shower cloud which can bring thunder and hail (see plates 14 and 15). Both of these types tend to be of large vertical extent, often being several kilometres deep from base to top.

The colour plate section shows examples of the main types of cloud. Remember though that due to the infinite variety of cloud formations, there will be plenty of occasions when the clouds you see in the sky look nothing like these pictures. It takes some degree of experience to be able to classify every cloud in the sky, and even two relatively experienced meteorological observers may not agree exactly on the classification of a particular cloudscape. On many occasions there will be more than one cloud type in the sky at the same time, making cloud classification even harder. One of the best ways to learn the appearance of the various types of cloud is to visit some of the excellent websites on this topic – see Taking it further for details.

BEAUFORT CLASSIFIES THE WIND

An instrument to measure wind speed proved remarkably difficult to perfect. The first system for classifying wind speed, developed by **Admiral Francis Beaufort** of the Royal Navy in 1806, was based on observing the effect of the wind on the ocean surface. This scale has been revised since to incorporate wind observations over land. The wind speed is classified into 13 'forces', with 0 corresponding to flat calm and 12 corresponding to hurricane-force winds causing extensive damage to buildings. Meteorologists still like to measure the wind speed in knots, which is another hangover from the nautical origins of wind measurements, with one knot corresponding to one nautical mile per hour. Most media weather forecasts these days give the wind speed in standard miles per hour (or in kilometres per hour in countries which use the metric system for measuring distance).

Force (mph)	Description	Equivalent mean wind speed measured at 10 metres above the ground
0	Calm. Smoke rises vertically.	0
1	Light air. Wind direction shown by smoke drift.	1–3
2	Light breeze. Wind felt on face. Leaves rustle.	4–7
3	Gentle breeze. Leaves and small twigs move. Flags lift.	8–12
4	Moderate breeze. Dust and loose paper lift. Small branches move.	13–18
5	Fresh breeze. Small trees sway. Crested wavelets on lakes.	19–24
6	Strong breeze. Large branches sway, telegraph wires whistle, umbrellas difficult to use.	25–31
7	Near gale. Whole trees move. Hard to walk against.	32–38
8	Gale. Small twigs break off. Impedes walking.	39–46
9	Strong gale. Slight structural damage (slates removed from roofs).	47–54
10	Storm. Seldom experienced land. Considerable structural damage, trees uprooted. on	55–63
11	Violent storm. Rarely experienced. Widespread damage.	64–72
12	Hurricane. At sea, visibility is badly affected by blowing spray, sea surface completely white.	>72

Table 7.1 The Beaufort Scale (land version).

Measuring the upper atmosphere

Up until the early twentieth century most weather observations were limited to measurements taken at or near the surface. However, knowing what the atmosphere is doing up aloft is crucial to understanding the developing weather and forecasting what is going to happen next. One of the main scientific experiments carried out by early balloonists was the recording of meteorological data, but it wasn't until the 1920s that a package of instruments could be attached to an unmanned balloon filled with helium or hydrogen and the readings taken could be transmitted back to the Earth's surface as a radio signal. These balloon-borne instrument packages are known as *radiosondes*, and they still play a key role in the gathering of weather data across the globe. During World War II, there was a massive increase in the numbers of aircraft flying at higher and higher altitudes and requiring more detailed weather information aloft. Consequently radiosonde technology developed rapidly and a network of observing stations grew up, many of which continued to provide data during peacetime.

RADAR – ANOTHER WARTIME DEVELOPMENT

Warfare during World War II was heavily reliant on long-range aircraft making bombing raids. In order to get some advanced warning of such raids, the British worked on the development of a system to detect approaching aircraft before they were visible to the naked eye. The result was RADAR – an acronym standing for *RAdio Detection And Ranging*. The principle was similar to the way that a bat senses its surroundings. A radio signal was emitted from a transmitter in a series of pulses which then reflected off any obstructions in their path and returned to a receiver next to the transmitter. By measuring the time taken for the pulses to return to the receiver, the distance to the obstruction could be calculated. If the pulse was transmitted horizontally, it would reflect off surface obstructions such as trees, buildings and hills, but by angling the beam up slightly, it could be used to detect objects in the air – such as approaching aircraft.

Radar operators soon noticed that, if the beam passed through areas of rain, the raindrops would act to reflect the beam, resulting in a cloud on their radar screens that obscured any signal from aircraft. To them this was just a nuisance, but after the war, a small group of scientists working at the British Government's Radar Research Establishment at Malvern began to experiment with radar as a meteorological observing tool. Careful selection of the radio pulse frequency allowed them to ascertain the size of the raindrops reflecting the beam, and hence calculate rainfall rates, not just at point locations like a rain gauge, but over a wide area. The usefulness of this development to weather forecasters cannot be overemphasized. For the first time, forecasters could watch the size and intensity of approaching areas of rain and see showers developing and decaying over regions with no weather observers on the ground. Plate 19 shows an example of a rainfall radar picture, for the UK on the 19 June 2007. The bright orange, red and pink colours indicate very heavy rain. On this day there was considerable localized flash flooding in parts of the country. Further work on rainfall radar technology, much of it in the States, has led to the development of the *doppler radar* which can sense not only the intensity of the precipitation but also the direction that the raindrops are being blown by the wind. This means that rotation in thunderstorms can be spotted. This rotation may be followed by the development of a tornado, and so doppler radars (which can even be mounted on vehicles) are vital tools in the issuing of warnings of potential tornado activity.

WEATHER SATELLITES – THE VIEW FROM SPACE

One further technological development in the twentieth century that was not originally designed for meteorological applications has had perhaps the most profound effect on weather observing and forecasting. On 4 October 1957, the Soviet Union launched Sputnik 1, the first artificial satellite to be placed into orbit around the Earth. This satellite transmitted a simple radio signal, but it wasn't long before meteorologists realized that satellites could provide an ideal platform for watching the atmosphere from above. On 1 April 1960, the US space agency launched their TIROS 1 satellite, which carried a television camera and a transmitter that could beam the pictures back to Earth. Although the pictures were rather poor quality it was possible to make out patterns of clouds associated with weather systems – the first time that such patterns had been observed from far above. This was a wonderful tool for meteorologists who were able to obtain a large-scale view of the state of the atmosphere without having to piece a picture of the weather together from sometimes rather widely spread surface observations. TIROS 1 didn't last long (the batteries failed after 11 weeks) but the age of weather observation from satellite was here to stay.

The development of weather observing networks

Single weather observations from a ground-based weather station are of limited use to weather forecasters. What forecasters really need is a network of observing stations across their area of interest so that they can build up a picture of the weather patterns across a wide area. By the latter part of the nineteenth century, an informal network of weather observing stations began to develop. In Britain and across the British Empire many of the most reliable records during this period were kept by clergymen who set up weather stations in their vicarage gardens. This was particularly true in India, where the clergy had good reason to record variations in the weather. During monsoon droughts, when food production fell dramatically, the local clergy were often the first people to initiate famine relief measures in their parishes before the state government could react.

Ports were also major observing locations, with sailors wanting to know something about the weather conditions before setting out to sea. The first true operational weather observing network was set up around the main ports of Britain in the 1850s by Admiral FitzRoy who wanted to be able to provide more information on the weather to shipping (see Chapter 8 for more details). This development was greatly aided by the invention of the electric telegraph, which allowed weather information from ports all around the UK to be rapidly relayed to a central office where it could be collated, plotted and used to draw up simple weather forecasts.

During the twentieth century, the growing importance of commercial and military aviation, and its dependence on the weather, meant that the network of operational weather observing stations started to shift away from ports towards airfields. Weather observations from airfields have a dual use. The first is the local use – pilots landing at an airfield can be informed of the prevailing conditions before they begin their approach, and can plan to divert to a different airfield if the conditions at their intended destination are too poor, most likely because of poor visibility or thunderstorms. On a wider scale, the observations can be passed to a central collection point where they can be plotted on charts and re-broadcast (these days by high-speed computer networks) to weather forecasters around the country or even further afield.

Observing the weather today

The collection of meteorological observations in the modern world is a complex and expensive business. Observations from land-based weather stations make up only a tiny fraction of the total amount of data collected on a routine basis, and the satellite pictures that we are used to seeing in media weather forecasts represent only a small proportion of the information that satellites can provide. Many of the observations used by forecasters to set the initial conditions for computer forecasts are obtained from instruments that record automatically all the time. Instrumented floats and buoys in the ocean beam their data back to the main meteorological centres via satellite link. Many commercial airliners also carry instruments that can measure winds and temperatures throughout the flight, which are transmitted via satellites. We'll now look at some of the major sources of observational data available to today's forecasters.

SURFACE OBSERVING STATIONS

There's a real mix of new technology and more traditional techniques when it comes to taking weather observations at surface stations.

Many weather stations these days are fully automated. Instruments have been designed which can sense pressure, temperature, humidity, wind speed and direction, rainfall rates and even cloud amounts and heights without any need for a person to read them. These instruments are very different from the original mercury-in-glass thermometers or mercury column barometers. Modern thermometers use the variations in resistance of a coil of platinum wire to sense the temperature. Barometers are based on a capsule from which most of the air has been evacuated, and which expands and contracts in response to changes in air pressure. Automatic rain gauges contain a small receptacle mounted on a pivot, which tips every time it fills up, flicking a switch. All these instruments can produce a signal of varying voltage, capacitance or current in an electric circuit, which can then be transmitted either down a wire or as a radio signal. Even at stations that are manned by a human observer, much of the weather data are sensed automatically and the observer then provides extra information on, for instance, cloud types and visibility. At some busy airports and military airfields, observations are made every 30 minutes, 24 hours a day, but hourly observations are more common at most operational stations. In addition to these main observing stations, many of which are at airfields, there are a large number of automated stations providing a more limited set of observations – perhaps just temperature and humidity – particularly around the road network. These are useful for forecasting weather conditions, and particularly the risk of ice, on the roads.

Despite all the many automated weather stations, even in the developed world there are still many observing sites that aren't so different from the stations set up by clergymen in the vicarage gardens in the eighteenth and nineteenth centuries. These stations, manned by volunteers, provide observations at most perhaps every three or six hours during daylight hours, but more often just once per day. They report maximum and minimum temperatures, rainfall totals and perhaps some additional information on winds and clouds. The instruments are sometimes provided and maintained by national meteorological organizations and the sites are inspected regularly to ensure that the observations are of sufficient quality. Schools, colleges and commercial organizations with an interest in the weather (such as water utility companies) are some of the main participants in this volunteer network, but some contributors are simply individuals with a passion for the weather.

As well as land-based observing stations, around 10,000 commercial ships carry observing equipment and provide regular data to National Meteorological Services. As well as providing standard weather readings, ships often report information on the sea state – the size, frequency and direction of waves and swell. There is also a growing network of moored weather buoys in the oceans that provide weather observations which are relayed by satellite to forecasters. The most extensive network is in the tropical Pacific, monitoring the state of the ocean and atmosphere in order to help with the prediction of El Niño events.

UPPER AIR OBSERVING

Observing the weather requires much more than observations taken at the Earth's surface. Much of the weather we experience at the surface is a result of developments several kilometres above the surface, and so to forecast these developments we need to know what is happening up aloft. One of the best tools we have for this task is still the radiosonde – a package of instruments borne aloft by a balloon filled with hydrogen or helium. A modern radiosonde weighs only a few hundred grams and can provide highly accurate measurements of temperature, humidity and pressure every few metres in the vertical. Depending on circumstances, some sondes even rise into the stratosphere before the balloon bursts and the instrument package floats back down to Earth on a parachute. Wind speed and direction can also be calculated, either by tracking the balloon by radar or using the Global Positioning Satellite (GPS) system to track the balloon's motion as it rises. Around 600 stations around the world release balloons twice a day, at 00 and 12 UTC. The vast majority of these stations are in North America, Europe, China, Japan and India leaving much of the globe very sparsely covered. In particular, apart from a few island stations in the Pacific, there is almost no radiosonde information over the oceans, where many of the weather systems develop that then go on to affect land areas.

Extra information on upper-level temperatures and winds is provided by commercial aircraft. These carry instrument packages that automatically record data along the flight track and transmit back to meteorological centres by satellite relay. By far the densest coverage of these reports is over the USA, due to the huge number of commercial flights there, but useful extra information is also provided by intercontinental flights. Because most commercial aircraft fly at altitudes of around 10,000 metres (30,000 feet), the data provided by them are largely restricted to this level. However, this often coincides with the level of the jet streams in the upper troposphere, which have a key role in the development of weather systems. Hence this added information is often vital in determining the exact strength and location of the jet streams at the start of a computer forecast.

MODERN WEATHER SATELLITES

Today's satellites have much more sophisticated equipment aboard than the original TV cameras, and provide an array of high-quality images to forecasters. They also carry instruments that don't just provide pictures, but actually make measurements of temperature, humidity, winds and even concentrations of gases such as ozone throughout the depth of the atmosphere. Some new satellite instruments include radar systems that look down into the clouds beneath them, providing information on the vertical structure of clouds which was previously unavailable to forecasters unless the clouds happened to be over a radiosonde station.

The satellites that provide pictures of cloud patterns fall into two categories. First there are the *geostationary* satellites. These are placed in orbit above the equator at such an altitude that they orbit the Earth at exactly the same rate as the Earth is rotating. Hence the satellites remain over the same point on the Earth's surface. This altitude is about 36,000 kilometres (22,400 miles) above the surface, almost three times the diameter of the Earth. Even so, the scanning equipment they carry allows them to provide very detailed pictures every 30 minutes. A geostationary satellite can provide a view of the whole Earth disc beneath it. However, because the satellites are positioned over the equator, they provide a rather poor view of the weather at high latitudes (polewards of about 60 degrees north and south) due to the curvature of the Earth. A network of five satellites evenly spaced around the Earth provide pictures which cover the whole planet. Two European satellites cover the Atlantic, African and Indian sectors. A Japanese satellite covers East Asia and the West Pacific, and two US satellites cover the central and east Pacific and the Americas.

Secondly there are *polar-orbiting*, or *sun-synchronous* satellites. These satellites are in a much lower orbit than the geostationary ones (about 850 kilometres or 530 miles above the surface), and consequently provide much more detailed images. They orbit over the Earth's poles and take about 100 minutes to complete an orbit. As they orbit, the Earth is rotating beneath them and their orbit is timed so that in any 24-hour period the satellite will have passed over every point on the Earth's surface twice – once on its *ascending* path as it tracks from South Pole to North Pole, and once on the *descending* path from North to South Pole (see Figure 7.1). Two such satellites, operated by the US National Oceanic and Atmospheric Administration (NOAA), are in operation at any one time. Because of the orbit over the poles, these satellites provide a much more detailed view of the weather patterns in the polar regions than the geostationary satellites. They scan a 'swath' beneath them, scanning from side to side across their track.

The images that both geostationary and polar-orbiting satellites provide are becoming more sophisticated all the time. There are three main types of satellite image, based on viewing the Earth at different wavelengths of electromagnetic radiation.

1 **Visible images** – *see plate 20. These are produced by the satellite sensing the sunlight reflected from the clouds and the Earth's surface, in much the same way that our eyes would do*

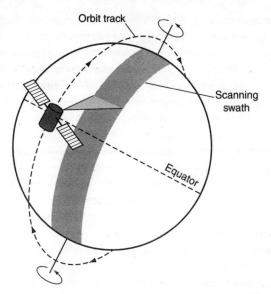

Figure 7.1 A polar-orbiting satellite scans a swath of the Earth beneath it as it orbits.

if we were sitting on the satellite. Clouds, and particularly thick clouds made up of water rather than ice, are particularly strong reflectors and so show up very bright in visible images. Fog also shows up well as does lying snow and ice. The surface of the ocean on the other hand absorbs much of the sunlight that falls on it and so tends to appear dark on visible images. The major drawback with visible images is that they rely on light from the Sun. During the hours of darkness no images are available.

2 **Infrared images** – *see plate 21. An infrared sensor detects radiation emitted by the objects it is looking at. Due to the temperature of the Earth and its atmosphere, the bulk of the energy radiated by the surface and clouds falls in the infrared part of the spectrum, and is an indication of the temperature of the object. So an infrared satellite picture is really a map of the temperature of the clouds and the Earth's surface. Because the ground is comparatively warm it appears dark in an infrared image, with warmer surfaces such as deserts being almost black, and cooler surfaces showing up as shades of dark grey. Very high clouds such as cirrus are very cold, and so appear bright*

white in the picture. Lower clouds take on various shades of grey, depending on the temperature of their tops. Very low cloud is often at a very similar temperature to the surface of the Earth just below it, and so is sometimes rather difficult to make out.

3 **Water vapour images** – *see plate 22. These images are very different to the visible and infrared pictures which appear frequently in media weather forecasts. Instead of showing the locations of clouds, water vapour images are a map of the temperature of water vapour in the atmosphere. Even parts of the atmosphere which are completely free of cloud will still contain water vapour, which emits radiation very strongly at certain wavelengths. By sensing this radiation, the satellite sensor can plot a map of the temperature of water vapour in the atmosphere. Regions where there is water vapour high in the troposphere will appear bright white. If the upper atmosphere is very dry then the satellite will detect water vapour in the lower troposphere and is therefore warmer. This will appear black on the image. The surface of the Earth is never visible in a water vapour image since even over the deserts there is still some water vapour in the atmosphere. The resulting picture is a rather ghostly image that makes the Earth look rather like a different planet such as Jupiter. The swirls and streaks on the image can be used by an experienced forecaster to ascertain the position and shape of features like jet streams, and careful interpretation of the image can help identify places where weather systems are developing rapidly.*

A combination of these three types of satellite image can be used to build up a detailed picture of the structure of weather systems. These pictures are particularly useful if a sequence of images is available, preferably as an animation. By noting places where cloud is forming or dissipating rapidly, or the patterns of clouds are changing, the development of the atmosphere can be tracked in close to real time, and short-range forecasts can be made by extrapolating the developments which can be seen in the sequence of pictures. A depression or frontal system in which the clouds are becoming brighter with time in a sequence of infrared images suggests that the cloud tops are getting higher and the clouds are therefore becoming thicker. This in turn suggests that any rain associated with the system will be getting heavier with time as the system develops. The appearance of convective clouds can be tracked in satellite images, indicating likely locations of showers. Once again, if the clouds become brighter with time on a sequence of infrared images this suggests that the showers will be getting heavier. The locations of areas of fog will be clearly shown in a visible satellite image as hard-edged, bright patches. Fog over hilly terrain sometimes reflects the positions of the valleys, with the hill-tops remaining clear of fog. Satellite images are useful for identifying patchy fog, particularly if the fog doesn't happen to coincide with the locations of surface-observing stations.

Vast amounts of other types of data can also be obtained from satellites. In fact, of all the observational data that go into setting the starting conditions for a computer forecast of the weather, at least 75 per cent comes from satellites rather than ground-based observations. Satellites can provide estimates of winds, either by tracking the motions of clouds at particular levels in the atmosphere or, over the oceans, by observing the scattering of light by waves, which will be proportional to the strength of the surface wind. Satellites can also provide vertical profiles of temperature and humidity in the atmosphere although these have rather coarse vertical resolution when compared to radiosondes launched from the surface. However, satellites allow near-global coverage, whereas the ground-launched radiosondes only provide information for very limited parts of the globe.

Making your own observations

Becoming your own weather observer has never been easier. A wide range of relatively cheap instruments is now available which will record many aspects of the weather automatically. Many high-street electronics stores sell simple 'weather stations' with sensors to record temperature, pressure and humidity which transmit data via a wireless link to a base station indoors. These are great for knowing what the temperature outside is like before you even get out of bed in the morning. More sophisticated systems are available from specialist suppliers over the internet, with automated sensors for most weather variables that can be linked to a home computer or data logger to make continuous records.

If you're out and about, it is even possible to purchase hand-held weather stations which are about the size and weight of a mobile phone. These instruments usually contain sensors for temperature and humidity, and a pressure sensor (barometer). The more expensive models also contain a small fan device which is used to measure wind speed. The readings from all of these instruments must be treated with a degree of caution as they will be subject to large errors due to problems with correct exposure. However, it can be very interesting and informative to look at how the various elements change during the period of a day's walk. However, remember that, if you are using a barometer on a walk in the hills, any changes in pressure will be dominated by changes in altitude rather than changes due to the movement of weather systems. As a general rule, the pressure will change by about one millibar (one hectopascal) for every ten metres (30 feet) of altitude. So the pressure you measure at the top of a 3,000 metre- (10,000 foot-) mountain will be about 700 millibars, compared to 1,000 millibars at sea level. Unless you are able to fix your altitude very precisely when taking pressure readings, any corrections you make due to your altitude will be subject to errors that may be larger than the true change in atmospheric pressure. If, however, you are in flat countryside, or on a boat at sea, the pressure changes your barometer records will be due to true changes in the atmosphere. In these cases, falling pressure will generally be a bad sign, signalling a potential deterioration in the weather, as discussed in Chapters 3 and 4.

Having your own weather records is a useful aid to producing your own weather forecasts. It is only by knowing what is 'normal' weather and what constitutes an extreme event that you can really start to make your own meaningful forecasts, and in Chapter 9 we'll go into the details of how you might go about doing this. By having your own instruments you'll start to get a feel for whether the forecasts you see on TV or the internet systematically over- or underestimate the temperature for your own location, and whether the wind speeds and directions in the forecast match up with what you actually observe. You can then try to explain systematic differences by looking at the local effects that might be causing the discrepancies. Is your garden in a frost hollow so that the actual minimum temperatures you record are always several degrees lower than the forecast? Or perhaps an exposed location on a hill means that the wind speed is always higher than forecast. Maybe the differences from the forecast are always greatest in a particular wind direction.

By having your own records, you'll be able, over time, to build up a picture of the typical weather conditions for your own site. Of course, it will take many years of observing before you can build up a truly representative record for a given location. The 'average' conditions quoted for individual stations in most books and weather reports are usually based on 30 years of data. However, a few years or even just months of data can help you to understand your own local weather conditions. You may also be able to pool your records with other weather enthusiasts in your local area, some of whom may have records going back many years. Many national meteorological agencies and societies maintain informal networks of observers who share their data through regular publications or the internet, so it is worth making enquiries and joining your national network. Most weather enthusiasts are only too pleased to share their observations with other like-minded people. Just occasionally weather observations taken by amateur meteorologists can even hold the key to understanding particular extreme or unusual weather events, so your observations may even end up helping the development of meteorological science.

Summary

Observations of the weather are a crucial first step towards producing a weather forecast. It is only by knowing what the weather is now, both at our own location and in the surrounding region, that we can start to extrapolate the current conditions into the future. Huge advances in technology over the past 100 years mean that we now have a vast amount of observational data which can be used to provide the starting conditions for producing weather forecasts. Of course, much of this information is only available to the large national meteorological agencies, but more and more is finding its way into the public domain in near real-time through the internet. See Taking it further for a list of various websites which provide good sources of satellite pictures and other types of weather observations. In the next chapter we will look in detail at how the large weather forecasting centres produce their forecasts, and how the observational data are processed to produce the best possible starting conditions for making these forecasts.

10 THINGS TO REMEMBER

1 *Man has always been interested in variations in the weather, and many ancient historical documents contain references to weather events and how they affected farming, travel and military campaigns.*

2 *The invention, during the seventeenth century, of both the thermometer and the barometer allowed weather observations to become quantitative for the first time.*

3 *The naming convention for clouds that we use today was devised in the early nineteenth century by an English chemist called Luke Howard.*

4 *Before wind speed could be measured directly by a scientific instrument, Admiral Francis Beaufort of the Royal Navy devised a wind classification based on the effect of the wind on the ocean surface and objects such as trees and buildings. The Beaufort wind scale is still used today, particularly for nautical weather forecasting.*

5 *The invention of radar and satellites, neither of which were originally conceived as weather observing tools, has given us a detailed insight into the formation of a wide range of weather systems which was previously inconceivable to meteorologists.*

6 *Geostationary satellites sit above the equator and orbit at the same speed that the Earth rotates. This allows them to provide a constant picture of the developing weather situation over one part of the globe.*

7 *Polar orbiting satellites are in a much lower orbit than geostationary satellites and orbit over the poles. They provide a very detailed picture of cloud patterns and pass over every point on the Earth's surface twice a day.*

8 Radiosondes, *which are a package of meteorological instruments mounted beneath a balloon, provide meteorologists with detailed measurements of the vertical variations in temperature, humidity and winds. This information is invaluable for weather forecasting.*

9 *A global network of manned and automatic weather stations, moored and drifting buoys, ships and aircraft operates 24 hours a day to provide observations to weather forecasters across the globe.*

10 *A relatively small investment in a set of weather instruments or an automated weather station connected to a PC allows all of us to become weather observers, checking the forecasts, sharing our data and even producing our own forecasts.*

8

Operational weather forecasting

In this chapter you will learn:
- *about the origins of operational weather in the nineteenth century to the present day*
- *how major weather forecasting centres produce their forecasts*
- *how to use and interpret weather forecasts from television, radio and the internet.*

If the weather affects our lives in any way, we'd like to know in advance what it has in store for us. This might be a matter of life-or-death importance; a single-handed yachtsman on a small craft wants to know if there's a risk of sailing into a storm. There might be financial considerations; a period of rain at the wrong time could ruin a harvest. Or the reasons might be as simple as wanting to hang the washing on the line or take a trip to the beach. For most of us, if we want to know what the weather will do, we can find a forecast on the television or radio, call one up on the internet or even get one sent to our mobile phone. The fact is that very few people produce their own weather forecasts without any reference to the information provided by state-owned or commercial weather forecasting agencies. Although it's a matter of global obsession to criticize the quality of the weather forecasts provided by the professional forecasting bodies, the fact is that no one working purely on their own can do any better. There will always be people who claim that they can, but any objective comparison of their forecasts with those provided by official sources will quickly disprove them. Of course, the forecast sometimes goes wrong, but taken over a reasonable length of time, the weather forecast provided by the main forecasting centres is mostly right most of the time. We'd be pretty foolish to think that we can beat the main forecasting centres at their own game. But we don't have to just take their forecast at face value. There are two main ways that we can take an 'official' weather forecast and make more of it.

▶ *The first is to add local knowledge. The forecast will generally get the broad picture correct, at least over the first 48 hours or so of the forecast period. Weather systems will generally be in about the right place. Areas of rain or strong winds will be there or thereabouts. Where we can score is by taking the forecast and applying some knowledge of how the local geography affects the weather where we are in a way that the forecasts which appear in the media just don't have time and space for.*

▶ *The second way in which we can add some value to a weather forecast is by recognizing where the potential for errors in the forecast are, and planning accordingly. The band of rain in the forecast might move slightly faster or slower than forecast, meaning that the timing of the rain might be different to the forecast. Or the intensity of the rain might be lighter or heavier. If we have access to a rainfall radar network via the internet (and in many countries this information is readily available – see Taking it further) then we can actually monitor the progress of the rain, see whether it is keeping to the forecast script or not and modify our expectations accordingly. This is what professional forecasters are doing all the time, and if we can pick up some of their skills and apply them to our specific circumstances then we can stay ahead of the game.*

If we really want a weather prediction, our first port of call has got to be the forecast produced by one of the main forecasting agencies and provided through one of the many sources available to us. So in this chapter we'll take a look at how these forecasts are actually produced. Only by knowing something about this process will we be able to form an opinion of whether the forecast is likely to be right or not, and where the main uncertainties lie. Then we can start to think about how we ourselves can add value to the forecast, which we'll do in Chapter 9.

A brief history

Routinely producing weather forecasts several times a day for broadcast through the media to the general public and a whole host of paying customers is a multi-million-pound business. Forecasts are produced on some of the world's most powerful computers (see Figure 8.1). A network of observing stations, both manned and automated, provide data in real-time from all around the globe. Satellites costing millions of dollars provide observations of weather systems from space. All of this effort is underpinned by an ongoing programme of research. But weather forecasting wasn't always like this. Before taking a close look at how forecasts are produced today, it is fascinating to look at how we got to where we are now.

Figure 8.1 At the heart of any modern weather forecast is a powerful supercomputer.

The first attempts to systematically predict the weather began in Britain in the mid-nineteenth century. At that time, Britain was at the heart of an empire built as much on international trade as military power. Ships left British ports daily for all corners of the globe carrying people and manufactured goods out to the various colonies. The returning ships brought raw materials to be processed in the factories and mills of Britain. The ships really were what kept the economy of the Empire going, and the loss of any ships due to bad weather or high seas meant a negative entry on the balance sheets, not to mention the loss of life. Hence it was the British *Board of Trade* who started the process that would eventually lead to the routine production of weather forecasts. In 1854 The Board of Trade appointed *Admiral Robert FitzRoy* as its 'Meteorological Statist' (these days we would probably say 'statistician') with the brief of providing an atlas of the winds around the world, using information collected by shipping as it travelled around the global trade routes. FitzRoy was already a celebrated figure in maritime circles for his mapping and exploration voyages. His charts of the waters around the coasts of South America were quite superb and remained the standard guides to vessels in this area for many years. He is of course also famous for being the captain of *HMS Beagle*, the naval survey vessel that carried the naturalist Charles Darwin on its voyages around the globe as he made the observations that would lead to his development of the theory of evolution.

FitzRoy set about his task, but wasn't content with simply recording and mapping the typical winds around the world. Regular loss of ships in UK waters set FitzRoy thinking about whether something more practical could be done to warn shipping of impending bad weather. FitzRoy believed that he could apply scientific principles to the atmosphere and produce forecasts of the weather that might be able to provide warnings to ships at sea or about to leave port. In fact it was probably FitzRoy who actually *coined* the term 'weather forecast'. But in order to produce forecasts, he needed to have information on what the weather was doing all around the country, and he needed it in real-time, not a day or two after it had happened. Twenty years previously this would not have been possible, but the invention in the 1840s of the electric telegraph meant that FitzRoy could get data rapidly from weather observers around the country who could transmit their observations back to a central office. This information could then be plotted onto a chart so that an overall picture of the weather around Britain could be rapidly established. FitzRoy was provided with a rather meagre staff of three people to help him with this task in the Meteorological Department of the Board of Trade – the forerunner of the Meteorological Office. FitzRoy recognized that changes in atmospheric pressure were a clue to what was happening to the weather, so he developed a barometer, or 'weather glass' with which to measure its fluctuations, and each port around the country was provided with one, together with a manual of how to interpret the pressure changes. These barometers were set up in prominent positions by the quayside so that every captain could have access to the readings before leaving port. Back at FitzRoy's HQ the pressure pattern and its tendency (rising or falling) was plotted, and forecasts were made on the basis of changes in pressure combined with knowledge of what had happened in previous similar situations. Forecasts could then be telegraphed back to the Port Authorities and warnings of impending storms could be posted to shipping.

In 1861, FitzRoy felt confident enough to issue his forecasts to the general public, and on 1 August in that year what was probably the world's first official, public weather forecast appeared in *The Times*. The text was short and to the point. It read simply:

> *General weather probable in the next two days in –*
> *North – moderate westerly wind; fine*
> *West – moderate south-westerly; fine*
> *South – fresh westerly; fine.*

In terms of scientific and technological advancement it probably doesn't quite rank up there with the apple falling on Newton's head or the Wright Brothers' first flight. But make no mistake, this was the start of something which has continued to hold a fascination for the public over the intervening years.

Unfortunately for FitzRoy, his forecasts were sometimes wrong, and by publishing them he was laying himself and his methods open to public scrutiny. In a foretaste of what would be the experience of every weather forecaster since, the public were often highly critical of the forecasts. Critical articles about the forecasts were published and the correctness of FitzRoy's methods was debated at the Royal Society. FitzRoy was a serious-minded man, and took the criticisms very personally – possibly not the best personality trait for a weather forecaster in any day and age, but especially one working without any of the modern aids to forecasting, using untried methods and relying on a less-than-complete scientific understanding of the atmosphere. FitzRoy was still convinced he could make his forecasts work, but a combination of ill-health and depression not entirely related to the criticism of his forecasts led FitzRoy to take his own life on 30 April 1865. In 1866, the forecasts in *The Times* were withdrawn, although they did recommence a few years later.

THE BERGEN SCHOOL

The US National Weather Service was set up in 1870s, but for the next real advance in weather forecasting we have to move forward a few years to the early twentieth century and look to Norway. At the University in Bergen, on the Norwegian west coast, a small group of scientists working under **Wilhelm Bjerknes** were linking together detailed observations of the weather and the basic physical principles that govern how the atmosphere works. The location probably wasn't a coincidence – Bergen gets *a lot* of weather. Sitting slap bang at the eastern end of the Atlantic storm-track, there aren't many days a year when Bergen isn't experiencing the effects of a depression or a frontal system. Average monthly rainfall during the autumn and winter is around 250 millimetres (ten inches), and despite being at a latitude of 60 degrees north, even during the long summer days, average daily sunshine hours never get anywhere near double figures. Weather like that gave this group of scientists, known as the *Bergen School*, plenty of opportunity to make their observations.

By applying the laws of physics, Bjerknes and his colleagues realized that it should be possible to write down a set of mathematical equations that described how the atmosphere would evolve with time. Finding a way of solving these equations would allow them to make mathematical predictions of the weather. But therein lay a problem. Without computers and the advanced mathematical techniques available to today's forecasters, there *was* no way to solve these equations. Instead they developed a more pragmatic approach to the problem. Their weather observations had shown them that the weather in their part of the world often followed similar patterns. Changes in temperature, pressure and wind directions all seemed to be correlated with episodes of wet weather in a way that could be categorized. Perhaps their most lasting contribution to the field was prompted by the observation that large changes in weather variables like wind, pressure and humidity often happened quite abruptly, over a period of only a few hours, to be followed by rather small, gradual changes over the next few days. They realized that there must be quite sharp boundaries in the atmosphere between masses of air with very different characteristics, and that much of the cloud and rain that they observed occurred on or near these boundaries. They termed these boundaries *fronts*, with a warm front denoting a transition from colder to warmer, more humid air, and a cold front marking the opposite transition. This terminology is still in use today (see Chapter 4), and the way that the Bergen School drew weather maps has influenced the methods and thinking of virtually every weather forecaster since the 1920s. The fact that much of the important work by the Bergen School happened at the time when World War I was taking place has led to speculation that the concept of atmospheric fronts was influenced by the Western Front – the virtually permanent boundary between the two opposing armies. Whatever the influence, the name has stuck, and not just the name but the whole way of looking at weather systems. The figures shown here, reproduced from a 1922 paper by Wilhelm Bjerknes' son Jakob and his colleague Halvor Solberg, will be instantly recognizable to any weather forecaster, meteorology student or pretty well anyone who's spent a bit of time studying weather maps (see Figure 8.2). They show the stages in the development of a typical low pressure system, in both plan view and cross section. This detailed picture is quite remarkable when you realize that it was developed without the advantage of satellite images, radar and very few measurements taken very far above the Earth's surface. Jakob Bjerknes lived until 1975 and went on to work on many different aspects of meteorology and the interaction between the atmosphere and oceans. One can only imagine the massive sense of excitement and achievement he must have felt when the first satellite pictures of weather systems viewed from space showed that his early conceptual model was proved to be largely correct.

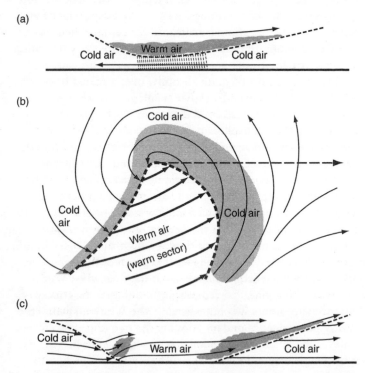

Figure 8.2 Diagram based on original figures from a 1922 paper by J. Bjerknes and H. Solberg showing cross sections and a plan view of a mid-latitude depression.

L. F. RICHARDSON

The Bergen School weren't the only people trying to apply the physical laws of fluid dynamics to the atmosphere to develop a mathematical method of forecasting. In Britain, **Lewis Fry Richardson** was also thinking along the same lines. He too realized that he wasn't going to be able to solve the equations exactly but he thought that, if he had a grid of meteorological observations across a geographical region, it might be possible to advance the equations forward in time in small steps at each point in the grid, using the information at the other points in order to do this. He also realized that this method would require thousands and thousands of calculations in order to make a forecast for any significant time into the future. This indeed was a daunting task, but Richardson broke the problem down into small steps. He devised a series of forms, each one of which held the details of a single calculation. Much of this work was performed during Richardson's service as an ambulance driver in World War I.

Insight – Richardson's first forecast

Using observations of winds and pressure taken across Europe on 20 May 1910, Richardson performed a six-hour forecast of the surface pressure. This took him six weeks and he double-checked every single calculation. Unfortunately the result was somewhat disheartening – he forecast a change in pressure of 145 hectopascals (millibars) in six hours. This result was two orders of magnitude larger than any observed pressure change over that period, and Richardson put the result down to errors in the initial conditions of the forecast. However, he was not put off completely and in 1922 published a book on his techniques called *Weather Prediction by Numerical Process*. In actual fact, the error was probably due to the fact that he was using a set of equations which allowed the possibility of unrealistic, high-frequency oscillations of a type not seen in the real atmosphere. Nowadays these oscillations are filtered out of the equations

(Contd)

by sophisticated mathematical techniques. However, because Richardson was the first person to try to forecast the weather in this way, he had no idea that such a possibility existed.

Richardson was in a very real sense a man ahead of his time. His numerical methods for forecasting the weather were absolutely made for computers, but the invention of a practical electronic computer was still 30 years away. Richardson envisaged a scenario where individual people would perform the calculations necessary to make a global weather forecast in a short enough time to stay ahead of the real atmosphere. He proposed that the only way to perform this task would be to have a large building full of clerks – he had in mind something like a large concert hall – all performing a particular calculation and passing their results to the next person in the sequence. All this activity would be directed by a 'conductor' who would highlight which set of calculations had to be done next. Richardson called this idea his 'forecast factory', but these days we can recognize what is basically the human equivalent of a supercomputer. In fact Richardson even referred to the clerks as 'computers'. Richardson's own estimate was that 64,000 people would be needed to perform a global weather forecast, but this was probably an underestimate – Peter Lynch of University College Dublin has recently reassessed Richardson's forecast and reckons that nearer 250,000 human 'computers' would have been needed to do the job while keeping ahead of the actual weather.

At the time that Richardson published his book on his numerical methods, the standard operational forecasting technique employed in the UK Meteorological Office was based on drawing a weather chart for the day in question and then looking back through a whole set of previous charts to find a situation that looked similar. The forecast would then be based on what happened in the earlier case. This was a very practical method, and easy to apply. However, since the atmosphere on any given day is always in a unique state, it was a flawed method. Why should the atmosphere follow exactly an earlier pattern? Richardson's method offered a scientific approach, based purely on applying the laws of physics to the atmosphere in order to predict the weather. While there were still many problems to be ironed out with the method itself and its practical application, (not least the invention of the electronic computer), Richardson had started the process that led to all modern weather forecasting as we understand it today.

WARTIME DEVELOPMENTS

World War II (1939–45) brought about several advances in operational weather forecasting. The use of long-range, high altitude aircraft for bombing and reconnaissance placed more demands on forecasters, who had to be able to predict winds at the levels that aircraft were flying, and determine whether they'd be able to see their targets when they reached them, and find their home airfields when they returned. The measurement of winds, temperature and humidity throughout the troposphere by instrument packages mounted on balloons, with data being radioed back to the ground became a widespread practice. This was accompanied by advances in understanding of how weather systems form and develop. It was becoming clear that what happened in the upper part of the troposphere had a strong influence on the weather at the surface. Forecasters such as **Reginald Sutcliffe** at the UK Meteorological Office started to devise forecasting techniques based on analysing the winds and temperature in the atmosphere at heights of about five kilometres above the ground. Forecasters could then identify the location of 'development areas' – places where weather systems might be expected to form and intensify (see Chapter 4). Long-range reconnaissance aircraft were also used to supply meteorological information to military planners. Perhaps the most critical example of this came with the D-Day landings, as the allied forces planned to invade northern France in June 1944. The success of the landings was critically dependent on good weather, and as the intended date approached the weather was far from settled. The allies' chief meteorologist, **Captain James Stagg**, advised a 24-hour delay to the landings, partly on the basis of weather reports from the crew of the aircraft which flew daily missions from airfields on the west coast of Britain over the Atlantic. They had found a potential 'weather window' – a spell of calmer weather that was heading in from the Atlantic towards Europe. Although the waters of the Channel were still choppy, the landings were achieved on perhaps the only suitable day that summer.

POST-WAR US – THE CHICAGO AND PRINCETON GROUPS

The idea, started by the Bergen School and L. F. Richardson, that mathematical methods could be used to forecast the weather, was also developing further during this period. One of Bjerknes' co-workers in Bergen, **Carl-Gustav Rossby**, had moved to Chicago University, and was working with a small team on refining the mathematical equations that could be used for this task. Meanwhile, in 1946, at Princeton University, a project was beginning which would make Richardson's idea a practical reality. Under the leadership of the mathematician **John von Neumann**, an electronic computer was being developed and weather forecasting was seen as an ideal problem to be addressed by such a machine. The ENIAC was the world's first programmable computer, using vacuum tubes and resistor technology. Various schemes were considered for the first numerical weather forecasts on this computer, and the one finally chosen was rather *simpler* than the set of equations that Richardson had used 30 years earlier. However, the results were rather more encouraging, and in 1950, the Princeton group published their findings, showing a reasonably realistic forecast. This was not a *weather* forecast in the sense that we would understand it today. It had no details of the locations of cloud, rain, temperature or winds at the Earth's surface. It was simply a 24-hour forecast of the atmospheric pressure at a height of about five kilometres above the surface. The forecast itself took about 24 hours to run on the ENIAC machine, and it would still have needed a skilled meteorologist to interpret it into a forecast of the actual weather. Nonetheless, it showed that forecasting by mathematical methods on a computer could work, and this is the direction that many of the major developments in weather forecasting have followed ever since. Richardson was among the first to offer his congratulations to the Princeton group, describing their work as 'an enormous scientific advance on the single and quite wrong result' which he himself had obtained 28 years earlier.

In 1954 an operational numerical weather prediction group was established in the US as a joint project between the Navy, Air Force and the civilian Weather Bureau. Operational forecasts started in May 1955, but they were actually beaten to it by the Swedish Military Weather Service, who began computerized operational forecasts in 1954. Their 24-hour forecasts took six and a half hours to produce. Only 40 minutes of this was actual computing time – the rest was taken up with manual preparation of the punched cards that contained the initial data that had to be fed into the machine!

Insight – The UK Met Office gets computerized

The UK Meteorological Office began experimenting with numerical weather forecasts on computer in the 1950s. There is something rather wonderfully British about their first excursions into this field. The Met Office didn't have their own computer, but they borrowed one that belonged to Messrs. J. Lyons, the tea and cakes company! They used it to develop a numerical model when it wasn't being used by Lyons to manage the deliveries of teacakes to their various corner house cafes. The next, more advanced computer that the Met Office was able to borrow was a Ferranti Mark 1 machine belonging to Manchester University. The University allowed the Met Office to use this machine during the night, when the staff and students had gone home. The Met Office dispatched a team of its brightest young staff to Manchester to do the computing work. They were lodged in a rather spartan guesthouse in Manchester and on occasions had to climb the University gates to get into the building to work when everyone else had gone home. The Met Office eventually found the money to invest in its own computer, and started operational numerical weather forecasting in 1965. In 1967 they ran the world's first computer forecast that actually predicted rainfall amounts rather than just pressure patterns.

Throughout the 1960s and1970s there were many advances in the science of numerical weather prediction, all helped by increasingly powerful computers. More and more atmospheric processes were being included within models, and forecasts were being made further into the future. In addition to the development of models and the computers they run on, more and more data from satellite and surface-based observing instruments have become available to feed into the models. In 1975 the European Centre for Medium-range Weather Forecasts (ECMWF), a joint European venture, was established in Reading, UK. The brief was to develop a weather forecasting model that could predict the weather up to ten days in advance. Operational forecasts started at ECMWF in 1979 and have continued ever since, with increasing accuracy. In 2006 the centre acquired a new computer – the most powerful in the UK at the time. In the US, the National Center for Environmental Prediction (NCEP) currently run a forecasting model twice a day which makes predictions up to 16 days.

This brings us up to the present day, so we'll now examine in a bit more detail how an operational weather forecasting centre goes about producing the numerical predictions which are the raw material of pretty much every weather forecast that gets printed, broadcast or uploaded.

Computer models

Ask people how weather forecasts are produced and most would say that a computer was involved in the process. But most would have little or no idea about what that computer was actually doing in order to produce those forecasts. And to most people that's fine – the computer model used by the forecasting agencies is essentially a 'black box', the contents of which they have no need to know anything about. However, if you really want to get the most out of the forecast, understand why it sometimes goes wrong, and use the information as the basis for producing forecasts tailored to your own needs, it helps to know a little about what goes on inside a model. It isn't necessary to know the details of every process and every line of code. Even within the agencies that run the models there isn't one single person who knows the intimate details of *every* part of the model. They are massive pieces of software, running to hundreds of thousands of lines of code and requiring some of the world's most powerful computers to produce the forecasts in a reasonable time. A general overview of what happens in a computer model, or *numerical weather prediction (NWP)* model as meteorologists call them, is all that's needed in order to start making some intelligent use of the forecasts they produce.

WHAT'S IN THE BOX? – THE BASIC COMPONENTS

At its heart, an NWP model is very simple. A few fundamental laws of physics describe how the winds and temperature of the atmosphere will change with time. These laws can be written down as equations and advanced forward in time to produce a forecast. Essentially that's what an NWP model is doing, at many points in the atmosphere simultaneously. However, there are several factors which make things much more complicated and go some way to explaining why weather forecasts will never be exact, and we'll look at these as they come up. Here's a summary of the basic physical laws that every NWP model must include.

Newton's Second Law of Motion
This law states that an object will remain at rest or moving at a constant speed unless acted upon by a force. Replace 'object' by 'parcel of air' in this statement and you have a law that governs how air parcels move, allowing the prediction of the winds. In order to apply this to the atmosphere, a list of the possible forces that could act on the atmosphere is needed. This list turns out to be pretty short. The fundamental force in the atmosphere that sets parcels in horizontal motion is the *pressure gradient force*, which acts to set air moving from regions of high atmospheric pressure towards lower pressure. Gravity also acts on the atmosphere, trying to attract parcels of air towards the Earth's surface. This force is

almost exactly balanced by the vertical pressure gradient force, which is trying to push air upwards from the higher pressure near the surface towards lower pressure higher in the atmosphere. The net result of these two forces is that the atmosphere neither collapses down to the Earth's surface nor flies off into space. The main way that the winds in the atmosphere are slowed down is by *friction* between the atmosphere and the Earth's surface, so this must also be included in the equation. Because the action of friction depends on the precise nature of the surface, and no NWP model can know the details of every single hill, building and tree, this is one of the complicating factors mentioned above, introducing inaccuracies into the forecast. Some corrections have to be applied to Newton's Second Law for the effects of the Earth's surface being curved and for the Coriolis effect of the Earth's rotation (see Chapter 3), but once these are included, three equations of motion can be written down for the atmosphere – one for motion in the east-west direction, one for motion in the north-south direction and one for vertical motion.

Mass conservation or 'continuity'

This is a very simple statement that mass can neither be created nor destroyed. For practical applications within the atmosphere, it means that any accumulation of air within a given volume due to the motion of the winds must be accompanied by a change of density, and hence the pressure of the air.

THE FIRST LAW OF THERMODYNAMICS

This law relates heat added to changes in temperature and pressure of air parcels. Here is another complicating factor. We know that virtually all the heat that gets into the atmosphere comes ultimately from the Sun. We have a very good idea of how much energy is arriving from the Sun at the top of the atmosphere at any given moment. However, it is much more difficult to determine exactly where in the atmosphere that energy will be converted into heat. This depends on the distribution of clouds, the nature of the Earth's surface, the amount of evaporation of water from the surface and the precise location in the atmosphere at which that evaporated water re-condenses, releasing its latent heat.

In addition to these fundamental laws of physics, any weather forecasting model also needs to keep track of water vapour in the atmosphere. The main source of water vapour is evaporation of water from the Earth's surface, particularly the oceans, and the main way of removing it is through rainfall. Once water vapour is in the atmosphere, it gets moved around by the winds, which is relatively easy to include in the model. It is much more difficult to predict exactly where that water vapour is going to condense and form cloud, and then how much of it is going to fall out as rain. A further complicating factor is that water in the atmosphere can be in either its vapour, liquid or solid form, and all three forms can exist in close proximity. Early weather forecasting models ignored water in the atmosphere altogether and just predicted winds and pressure patterns. It was left to the forecasters looking at the model predictions to decide where it might be cloudy or raining.

APPLIED PHYSICS

The laws of physics that determine how the atmosphere will change with time are actually pretty simple. If we can apply these laws at every point in the atmosphere, starting from a known *initial condition* of the atmosphere, we ought to be able to predict the weather. Weather forecasting models are trying to do just that. But the complicating factors such as friction at the Earth's surface and the formation of clouds make this task impossible to achieve *exactly*. Computer models include a whole host of routines to deal with these processes that aren't explicitly included in the main equations. These routines are often described as 'model physics' for reasons that aren't very obvious since everything that a weather forecasting model is trying to represent is essentially just applied physics. The processes that are usually included in this 'physics package' are:

▶ *The passage of energy in the form of radiation through the atmosphere, both in terms of the short-wave radiation from the Sun and long-wave radiation emitted by the Earth's surface and the atmosphere itself. This includes the absorption and reflection of radiation by clouds, the absorption and emission* of radiation by the greenhouse gases in the atmosphere and the effect of the ozone layer.

▶ The formation of clouds and precipitation, due either to large-scale vertical motion of air leading to cooling and condensation (such as at a weather front), or by convective motion.

▶ The exchange of heat, moisture and momentum between the atmosphere and the Earth's surface. This includes evaporation of water, either from the oceans or land surfaces, friction due to rough surfaces such as buildings, forests or mountains, or the direct heating or cooling of the air by contact with the Earth's surface.

▶ The dissipation of momentum in the atmosphere by waves generated by the flow of winds over mountain ranges. Such waves, called gravity waves by meteorologists, can break down into turbulent motion, as anyone who has ever experienced turbulence when flying in an aircraft near mountains will know.

▶ Most weather forecasting models also have schemes for working out what happens to rain once it falls onto land. Some rain will soak into the soil; some will run off into rivers and eventually the sea; and some will stay on the surface in puddles or sitting on the leaves of plants and trees. Why should this affect the weather forecast? In fact the amount of water on and in the ground can have a big impact on the weather – see Chapter 5 for more details. Wet soil is slower to warm up during the day than dry soil, so the maximum temperature on a day when there has been heavy rain overnight will be lower than on a day when it hasn't rained for many days. Maximum temperature records are usually only broken after a prolonged dry spell, when all the moisture in the top layers of the soil has evaporated. The ease with which convective clouds can form is also determined partly by how much water is available to the air parcels near the ground. There can be very intense convective air currents rising over desert regions due to the very hot surface, but clouds don't form because there isn't a source of water vapour.

BUILDING THE MODEL

The physical principles that determine how the atmosphere evolves are well established, and equations can be written down which allow their effects to be quantified and projected into the future. How does a weather forecasting model actually go about applying these principles?

The main way that models work is by dividing the atmosphere up into a lattice of *grid boxes*. Imagine drawing a grid of lines on the Earth's surface, much like the lines of latitude and longitude. Then imagine slicing the atmosphere up in the vertical too. This will result in a whole set of three-dimensional cubes. In the vertical, the *model levels* are shaped to fit over any mountains on the surface as shown in Figure 8.3. Simply using height above sea level as a vertical co-ordinate wouldn't work, because some points in mountainous regions would actually be underground. Depending on the model itself, the grid might extend right around the whole globe or may just cover a more limited area such as a single country or continent. The size of the grid boxes may also vary depending on the model. In the current global model used at the European Centre for Medium-range Weather Forecasts (ECMWF) the grid boxes are about 25 kilometres square in the horizontal, whereas in the model that the UK Met Office uses to produce forecasts for just the UK and surrounding areas, the grid boxes are four kilometres square. Each cube or grid box contains one value of each of the weather parameters being forecast – winds in all three directions, temperature, humidity, pressure and cloud amount. The model solves each of the governing equations simultaneously in each grid box. The information from the surrounding grid boxes is used to perform these calculations. For instance, if the wind in a certain grid box is blowing from the west, then the air that will be arriving in that box will have the characteristics of the box to its west. The forecast is advanced in small steps of time, called, not surprisingly, *time-steps*. Depending on the model, and the size of the grid boxes, the time-step can vary from perhaps only 15 seconds for a model with very small grid boxes, to maybe 15 minutes for a model with larger grid boxes.

There are basically two types of NWP models. *Global models* cover the whole globe, whereas *regional models* produce a forecast over a more limited area, but with more detail. The area covered by an NWP model is usually referred to as the *domain* of the model. Some weather forecasting centres run both types of model, whereas others run only global or only regional models. Both types of model have particular benefits and drawbacks, and are used for specific purposes. It is useful to know, if you're using a weather forecast which has been generated by an NWP model, which type of model it came from and what sort of information it will be able to provide you with.

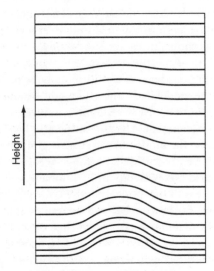

Figure 8.3 Arrangement of vertical levels in an NWP forecasting model.

GLOBAL MODELS

There are several reasons for running a global weather forecasting model, even if you happen to be most interested in the weather at one particular location. The first is simplicity. The real atmosphere is obviously continuous around the globe. Therefore it makes sense to build a model that doesn't have any artificial boundaries to deal with. Real weather systems don't disappear off the edge of the map – they carry on moving around the globe, so a global model is a more realistic representation of the real system. A second reason for choosing a global model is the length of the forecast period. The best way to illustrate this issue is with a practical example. Let's say that a particular forecasting centre wants to produce forecasts 48 hours into the future for western Europe. Over that time range it is unlikely that the weather around the other side of the globe, over Japan say, will have any influence on what happens over Europe. However, push the forecasts further into the future, and by seven days it is quite possible that weather conditions over the Pacific may start to influence European weather. Although individual weather systems don't move around the globe that fast, there are waves that can propagate through the atmosphere on this kind of timescale, causing weather systems to be generated as they go. So, if you want to produce forecasts for more than five days into the future, the weather anywhere on the globe might start to have an influence on the weather anywhere else, so it becomes important to use a model with a global domain. A third reason for using a global model is the requirements of the users of the forecasts. Most forecast users want to know the weather conditions at a specific location or over quite a small area. One major exception to this is international airlines, who need to know the winds and weather conditions along their flight tracks, covering a global network. It clearly makes sense to produce these forecasts using a global model rather than a whole set of regional models that might not exactly agree with each other at the boundaries. In fact the national weather services of the USA and the UK have a commitment, through the International Civil Aviation Organization, to produce global wind and weather forecasts for commercial airlines.

REGIONAL MODELS

Regional models are used to produce detailed weather forecasts over specific regions. These regions may be rather small – the Met Office runs a model that just covers the UK itself and its immediate surroundings; or rather large – the Met Office also runs a regional model that covers the whole of the North Atlantic and western Europe. The main benefit of regional models is that they can be run with smaller grid boxes than global models, and hence can provide more detailed and hopefully more accurate forecasts. Available computing power is always the main constraint on how small the grid boxes can be in a regional model. The smaller the boxes, the more are required to cover a particular region, and so more calculations are needed to produce the forecast. More calculations take longer to perform and so there comes a point when the forecast takes too long to produce to have any practical use. However, the rapid developments in computing power mean that regional models are becoming more detailed. At the beginning of the twenty-first century, a regional model covering the UK with grid boxes ten kilometre square would have been considered pretty cutting edge. Currently (2008) the Met Office runs a model covering the same UK area with grid boxes four kilometres square, and is testing a version in which the grid boxes are only one kilometre square, in readiness for the next upgrade to their computing resources.

A model with very small grid boxes can contain a lot of detail about the land surface conditions. Smaller hills will be resolved, headlands and bays around coastlines will be better defined, and the nature of the Earth's surface can also be specified in more detail. All of these things can affect the weather. In particular, convective showers are often triggered by air flowing over or around small irregularities in the underlying surface such as a small hill, or a headland. On 16 August 2004, a series of heavy showers were generated by a small peninsula on the Cornish coast of the UK (see Chapter 6 for more details). At the time the Met Office was using a forecast model with grid boxes twelve kilometres square. Although it had predicted showers in the area, the rainfall amounts were under-predicted and the precise locations of the heaviest showers were not captured by the forecast. Subsequent re-runs of the forecast using the four kilometre grid box model were much more successful at capturing both the intensity and the locations of the heaviest rain to the extent that it might have been possible to issue warnings of the impending flooding several hours before it actually occurred.

A major drawback of regional models is that, unlike the real atmosphere, they have edges. In reality, weather is still happening beyond these boundaries, and weather systems can travel into the model region from outside the boundaries. How do regional models deal with this issue? In fact regional models have to get help from global models – they can't exist in isolation. Prior to a regional model producing a forecast, a global model must have been run to produce *boundary conditions* for the regional model – that is information about what is happening at the edges of the regional model domain. A depression that develops outside the domain of a regional model might affect the region in the next 24 hours or so, so the regional model needs to be fed with information about the system as it arrives at the boundaries. In such a case, the global model that provides this information is known as the *driving model* since the information it produces drives the conditions at the regional model boundaries.

Most major forecasting centres therefore run a 'suite' of NWP models, which will include a global model and one or maybe more regional models covering smaller areas with more detail. Some national forecasting centres run a regional model that covers their own country and its immediate surroundings, but do not have the computing resources necessary to run their own global model. In this case they get their boundary conditions from one of the forecasting centres that does run a global model. At time of writing, the Met Office is running a suite of three models. Twice a day, soon after midnight and midday, a global model forecast is produced up to six days ahead. This model has grid boxes approximately 40 kilometres square, and 50 levels in the vertical. It produces boundary conditions that drive two regional models. One covers the whole North Atlantic and western Europe, with grid boxes 12 kilometres square and 38 vertical levels, and this model produces forecasts up to two days ahead. The other regional model just covers the UK and nearby parts of Europe, with grid boxes four kilometre square, and also has 38 levels in the vertical. It too forecasts up to two days ahead. One clever thing about this system is that the global model 'talks' to the regional models as it runs, providing boundary conditions that are used almost instantaneously by the regional models. That way the regional models don't even have to wait for the global model to finish before they can start running, meaning that the forecasts are produced several hours quicker than they would be if the models had to run one after the other. The two regional models actually run four times a day, with updates at 06:00 and 18:00 in addition to the main forecasts at 00:00 and 12:00. These runs also require boundary conditions from the global model, so the global model is also run at these times, but only for 48 hours and solely to produce the driving conditions for the regional models.

This may all seem like a lot of unnecessary technical detail. However, it helps to know what sort of model produced the weather forecast we might be trying to use. For instance, if we happen to be worried about the occurrence of heavy showers over a very small area, a rainfall forecast generated by a global model with grid boxes perhaps 50 kilometres square is unlikely to capture realistically heavy rainfall rates, and certainly won't be able to get the locations of showers in exactly the right places. This doesn't mean that the forecast is useless, it just means that we should apply caution when interpreting it. However, if we happen to be sailing a boat in a transatlantic race then the wind forecasts generated by a global model will probably be our best guide to the conditions we might expect en route, particularly when we start to look more than a few days in advance.

The process of forecasting

INITIAL CONDITIONS

NWP models work by applying the relevant laws of physics to the atmosphere and using these laws to advance the state of the atmosphere forward in time. This method assumes that it is possible to provide the model with information about the state of the atmosphere at the start of the forecast. This turns out to be a crucial part of the forecast process, if not the single most important part. Weather forecasting is what mathematicians would call an *initial value problem*. This means that the conditions that you put in at the beginning have a major influence on the answer that you get out at the end. So, if we want to run a global NWP model with hundreds of thousands of grid boxes, we need to start off with a value of all the forecast variables – winds, temperature, pressure, humidity, etc. – at every single grid box. Not only that, but we want those values to be as close to their actual values at those points in the atmosphere at the initial time as possible. Even with the wealth of information available to us from satellites and other sources (see Chapter 7) it is pretty clear that we won't have all the actual observed values of these quantities that we need. So what do we do? The answer is very simple and comes from the models themselves. All operational forecast centres are running their models every six or 12 hours, producing forecasts perhaps 48 hours or more into the future. So what the forecast centres do is save the short-range forecast from the previous run of the model to act as the basis for the initial conditions for the next forecast. So if the model runs every six hours, then the six-hour forecast is saved and used to start the next forecast. In one way this is an ideal solution because the forecast model will produce a six-hour forecast of exactly the right variables on exactly the right grid to act as the initial conditions for the next forecast. Not only does this give values in every grid box, but those values will all be physically consistent with each other since the model used the physical laws that govern the atmosphere to calculate those values. So the pressure will be consistent with the temperature pattern, and the winds will be in agreement with the pressure pattern. The problem with this method is that, even if the initial conditions for the previous forecast were exactly correct, by six hours into the forecast they will already have started to drift away from reality, and in fact they certainly weren't exactly correct in the first place. However, in the six hours that have passed since the previous forecast was made, many observations of the actual state of the atmosphere will have been made and so we can use these observations to help push the initial conditions for the next forecast back towards reality. How this is done is very complex, and encompasses a whole branch of operational weather forecasting known as *data assimilation*.

DATA ASSIMILATION

Data assimilation is all about trying to get the best possible initial conditions for every NWP forecast that is produced. There has to be an element of pragmatism involved because we know that we will never be able to specify perfectly what the state of the atmosphere is at any one time, and even if we could then the approximations and simplifications included in the NWP model would mean that the weather which the model predicts will still be different to reality. Data assimilation also has to accept that there will also be errors in the observations themselves that may also contribute to errors in the forecast.

The set of conditions provided by the previous forecast is known as the *background field*. If observations of particular variables are available in a particular grid box then it is not sufficient to simply replace the value from the background field with the observed value. This would result in the values of these variables not being physically consistent with their neighbours in the grid or with a different variable at the same point. For instance, the winds that blow in the atmosphere are a result of the pressure pattern, so putting a wind observation into the model without also adjusting the pressure would result in the winds and pressure being out of balance. If we started a numerical weather forecast from a set of initial conditions which weren't physically consistent in this way then the laws of physics within the model would take it upon themselves to try and even out the inconsistencies at the start of the forecast. The way that this would happen would be through the generation of large-amplitude, high-frequency waves in the atmosphere that would rapidly dominate over any real weather systems. The effect would be similar to throwing a stone into a pond of still water – waves would rapidly spread out from the point at which the water was disturbed. This is exactly what happened in L. F. Richardson's forecast for 1910 when he predicted a pressure change of 145 millibars in six hours. Unfortunately for him, since he was the first person to actually attempt a numerical weather forecast, no one knew anything about this particular pitfall. Thanks to much pioneering work from the 1930s onwards by people such as Carl-Gustav Rossby and Jules Charney, we now know that we have to be very careful to make sure that the initial conditions for any weather forecast are properly self-consistent.

Instead of simply replacing the background field with observed values, data assimilation uses the observational data to gently 'nudge' the background field towards the real state of the atmosphere (see Figure 8.4). Observations don't just affect a single point in the model grid, but instead are allowed to have an influence on all the surrounding points as well, and this influence decreases with distance away from the central point. Observations of variables that change fairly slowly and smoothly in space such as pressure can have an influence on grid boxes up to 1,000 kilometres away, whereas observations of variables which change more rapidly over short

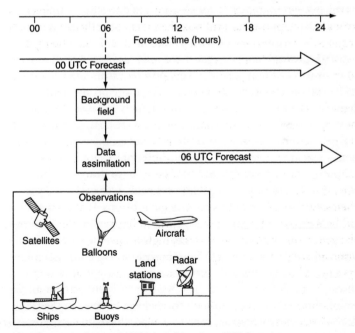

Figure 8.4 The process of generating initial conditions for an NWP forecast.

distances such as humidity will only have a more localized influence. If an observation of pressure is inserted into the model, the values of wind in the model around that point will also be adjusted in order to ensure that these variables remain consistent with each other.

Data assimilation also has to take account of the fact that all weather observations may have errors in them, due to the design of the instruments or the way the observations are taken. Some types of observation are more likely to be accurate than others, and this too has to be accounted for. If the known margin for error for a particular type of observation is large, it will be given less weight in setting the initial conditions than an observation with a smaller known error margin. Some observations may be ignored completely if their margin for error is actually larger than the difference between the observed value and the background field.

Data assimilation schemes are also designed to stop completely wrong observations from having an influence on the forecast. For instance, a weather observer on a ship out at sea may have reported a surface pressure reading that is ten millibars lower than any other nearby reading, and also very different from the background field. Ship observations are known to be more prone to errors than observations from land stations, largely because ships are moving, and so may be experiencing pressure changes simply due to movement across a pressure gradient. This means that observations from ships are treated with caution, and data assimilation schemes are programmed to reject them if they seem to disagree with other observations or the background field. This can lead to the occasional forecast going very wrong. If the observation turns out to be correct because there was perhaps a depression developing in the vicinity then this vital information will have been missed out and a brewing storm will not have been forecast. Data assimilation has to tread a very fine line between making sure that wrong observations don't influence the forecast and ensuring that all the useful information gets a chance to send the forecast in the right direction. Some forecasting centres actually have human forecasters monitoring the data assimilation process as it proceeds. As some of the observations are inserted into the model, they appear on a screen in front of the forecaster, along with an indication of whether the data assimilation process has taken account of them. The forecaster can then double-check observations which get rejected against other sources of data such as satellite imagery. If there is a cloud pattern that suggests that the pressure may indeed be falling very quickly then the forecaster can force the model to take account of the observation. This is done by *making up* phoney observations in the vicinity of the original one – a process known as *bogussing*. Faced with overwhelming evidence of maybe 10–20 observations all showing the same thing, the model has no choice but to accept it.

Insight – Bogussing saves the forecast

An example of the value of bogussing occurred in late December 1999. On 26 December, a deep area of low pressure swept across north-west Europe (see Chapter 6 for more details of this storm). Twenty-four hours earlier the duty forecaster at the UK Met Office had noticed that the data assimilation cycle was rejecting ship observations in the mid-Atlantic because the surface pressure being reported seemed too low. A careful check against the available satellite images showed that cloud was developing in the vicinity of these observations which had all the hallmarks of a rapidly developing low pressure system. So the forecaster decided to 'support' the rejected observations by inventing lots of similar ones in the same area. This was done and the forecast of the depression as it moved into Europe was rather good. However, the forecast made by some other forecasting centres, using the same observations, failed to capture the intensity of the depression and the strength of the winds.

RUNNING THE FORECAST

Once the initial conditions for the forecast have been generated by the data assimilation process, the forecast can be run on the computer. Compared to the original numerical forecasts performed in the 1950s when it took about 24 hours of computing time to produce a 24-hour forecast, the speed of forecast production these days is astounding. At the ECMWF, the ten-day global forecast takes about 75 minutes to run on the current supercomputer. Developments in computing power are moving at such a pace however that it won't be too long before this is considered somewhat slow!

Using an NWP forecast

Once the forecast has been run, maps of all the relevant variables can be produced – temperature, rainfall, winds, etc. One might think that this is the end of the process. The data can be used to produce all the graphics we're familiar with seeing on television or internet forecasts, or on the back page of the newspaper. But there's still a vital part of the forecasting process that has to be gone through. Just as a pile of bricks, beams and tiles doesn't constitute a house, the output from an NWP model is not a finished product, but rather the raw material. This is the point in the forecast process at which human forecasters take over from the computer and use their skill and experience to translate the numerical forecast into something useful for their customers and the general public. At most major weather forecasting centres, a small team of the most experienced forecasters will be studying the maps produced by the computer, considering how good the forecast is and putting together *guidance* for other users of the output from the NWP model. In some forecast centres this guidance may take the form of written text that accompanies the maps produced from the computer forecast. This text is often reinforced by face-to-face discussions or telephone conferences with other forecasters working at remote locations such as airfields. In the US, forecasters take part in a web-based discussion, which is actually posted on the internet in near real-time for anyone with an interest in the weather to read. Such discussions tend to be littered with jargon and technical terms such as 'trough disruption' or 'potential vorticity intrusion' which mean nothing to the average person in the street, and have limited meaning even to most amateur weather enthusiasts. However, this type of discussion is vital to make sure that the people in the chain of the forecast production process, from senior forecasters at the main centres to TV broadcast meteorologists and forecasters at airports and military air bases, all know what the main thrust of the weather story for that day is.

There's more to the process of interpreting the NWP forecast than simply translating the numbers from the computer into a forecast in everyday language. One of the most important roles of the professional human forecaster is assessing the output from the computer and deciding whether it is actually giving good guidance as to how the weather will pan out. But how does a forecaster know whether the forecast is going to be right or not before the weather has actually happened? There are a number of different ways that forecasters go about assessing the quality of a forecast. Let us say that a particular forecast has a start time of midnight on a particular day, and is a forecast for the next 48 hours. By the time the data assimilation process has been completed and the forecast has actually run, the actual time will probably around 3 or 4 a.m. This means that for the first three to four hours of the forecast, there are observations of how the real atmosphere has evolved through that period. An experienced forecaster can use those observations to make comparisons between the forecast and the real atmosphere during that short period. In particular, satellite and rainfall radar images will show the patterns of cloud and precipitation that can be compared with what the NWP forecast said was going to happen. As well as looking at general patterns of cloud and rain, the forecaster will also be looking for subtleties in the observations; places where clouds are getting thicker perhaps, or rain getting heavier. They can then look at the computer forecast to see whether these developments have been predicted. If the model seems to be handling the situation reasonably well through this early stage of the forecast then the forecaster can place more faith in its predictions further into the forecast period. If however there are already discrepancies between the forecast and reality at this stage, then the forecaster needs to start thinking about how these will knock on into the rest of the forecast. Figure 8.5 shows a senior forecaster at a national meteorological centre. An array of computer screens display output from the NWP model together with satellite pictures and charts of weather observations to help the forecaster assess the quality of the forecast.

Figure 8.5 A modern weather forecaster at work. Vast amounts of weather data are available on the screen at the click of a mouse.

An example of timely intervention by a human forecaster occurred in the UK on 30 January 2003. The NWP forecast that morning had predicted a band of largely light rain mixed with a little snow, extending across East Anglia for the evening rush-hour period. Looking at the observations during the morning, it was clear to the chief forecaster at the Met Office that there was actually a good possibility that the predicted light rain mixed with a few flakes of snow could actually turn out to be quite significant falls of snow. With 12 hours lead time the forecast was duly modified and snow warnings were issued. Unfortunately, because significant snow in this area is a fairly rare event, it took a while for the road gritting and snow clearance teams to swing into action, and many people missed or simply ignored the warnings. The heaviest snow coincided with the evening rush hour so that traffic ground to a standstill, blocking the roads and preventing the snow clearance and emergency vehicles from getting to the snowbound areas. Hundreds of motorists were stranded overnight in freezing conditions.

An experienced forecaster will also have a good knowledge of what are known as the *systematic errors* of the particular NWP model that they are using. In certain situations, most NWP models behave in ways which are wrong, but consistently so, and these types of error can be accounted for before the forecast is actually issued to customers and the public. This might be something as simple as knowing that, at a particular location, the model always under-predicts the minimum temperature by 2°C. The forecaster can simply add that two degrees back before issuing the forecast. A more complex example might be that, in a cold, unstable airflow during winter, the model does quite well at predicting showers over the ocean, but fails to capture the extent to which these showers will penetrate inland before dying out. In this case it takes real skill to spot the right conditions and then decide where the inland penetration of showers will actually occur. This isn't something you can learn from reading a book. It takes experience and discussion with other forecasters to start being able to make this type of adjustment to an NWP forecast.

Further into the forecast period there are clearly no observations to help the forecaster decide whether the forecast from the computer model will be right or not. At forecast ranges beyond 24 hours, the forecaster must rely more heavily on what the computer model is telling them. There are still ways of assessing the reliability of the forecast though. One method is to look at the forecasts produced by other computer models. While each major forecasting centre generally runs only one NWP model, there is an agreement between centres, brokered by the World Meteorological Organization (WMO), to share information, which includes observations and the output from computer models. With today's high-speed electronic communications networks, the output from a computer model from the US can be on the computer screen in a forecasting centre in Europe almost as quickly as it becomes available to the forecasters in the US, and vice versa. This means that forecasters will have the output from perhaps three to five different computer models. Each model is based on the same laws of physics and uses a similar set of observations to provide the initial conditions, but each one is formulated slightly differently. If all the models are going for pretty similar conditions through the forecast period then the forecasters can be more confident that the guidance they are providing is likely to be reasonably reliable. However, there will always be occasions when the different NWP forecasts all diverge and sometimes by only 24 hours into the forecast, they may all be going for very different conditions over a particular region. In these circumstances experience and judgement come into play again. Senior forecasters will have a good idea about which models are most reliable. It may not always be their own model, and it may vary depending on the particular weather conditions. It may be that three of four models are all producing a similar forecast, whereas one single model is out on a limb with a very different forecast. In this case the tendency is to go with the majority consensus, but not always.

NWP VERIFICATION AND DEVELOPMENT

The WMO co-ordinates an ongoing programme of NWP model comparison, producing performance statistics for the various models to help forecasters make informed choices about which model or models to follow. Naturally enough these statistics become the basis for informal competition between the various forecast centres. No one wants to see their model sitting at the bottom of the league table, and the forecasting centres near the top of the league want to maintain their position. The research and development teams at each centre are constantly striving to improve performance by modifying their model or making better use of the available observations to set the initial conditions. Sometimes quite small changes to the model can have an appreciable impact on the quality of the forecast. At other times, big expensive upgrades only have a minimal impact. Most forecast centres, as well as running their operational forecast, also run what is referred to as a *parallel suite*. This is an identical forecast, but using a version of the model that is acting as a test-bed for new developments. Only when a change to the model has been shown to produce an improvement (or at least no detriment) to the quality of the forecast over a period of time will it be included in the operational forecast model. Centres may also re-run forecasts from the past for cases when the operational forecast was known to have gone wrong. These so-called *hindcasts* will give an idea of whether the planned changes would have helped the model make a better job of a situation when it originally performed poorly.

Ensemble forecasting

As well as looking at the forecast provided by more than one NWP model, forecast centres are also increasingly using a technique based on running the same forecast, with the same model, many different times. The only things that differ are the *initial conditions* for the forecasts. This technique is called *ensemble forecasting*. The philosophy behind this idea is based on the chaotic nature of the atmosphere, and the fact that we cannot precisely measure the state of the atmosphere at every single point. As we have already seen, the outcome of any forecast is strongly dependent on the initial conditions with which we start the forecast. In some cases, changes to the initial conditions will have very little impact on the outcome of the forecast. At other times, seemingly insignificant changes to the initial conditions can send the forecast off in a completely different direction. It makes sense then to try to account for this when we produce a forecast. What is done in practice is that a range of initial conditions for the forecasts is generated by the data assimilation process. Each one is equally likely, and each one is equally in agreement with the observations that went into the assimilation process. The aim is to provide a set of initial conditions that will allow differences between the forecasts to grow with time. A group, or *ensemble*, of forecasts is then run using each of these initial conditions, with each forecast being referred to as an *ensemble member*. This then allows the human forecaster to assess the range of possible forecast outcomes. If, two or three days into the forecast period, all the ensemble members are still predicting very similar conditions, then the atmosphere is considered to be in a reasonably *predictable* state. If all the members diverge considerably through the forecast period, then the atmosphere is considered to be rather unpredictable, and we would have less confidence in the truth of the forecasts. What often happens in practice is that the forecasts tend to fall into groups, usually referred to as clusters. Perhaps 60 per cent of the ensemble members all give fairly similar conditions after three days. Maybe 30 per cent of the members then differ from the first set, but all agree reasonably well with each other. The other ten per cent give yet a third set of similar conditions. This is all useful information for the forecaster. The forecast that is actually issued is likely to be along the lines of the majority of the ensemble members, but with the knowledge that things might turn out differently. Forecasters in the media, particularly on TV and radio, will often try to convey some degree of this uncertainty when they deliver their forecasts. Sometimes this uncertainty might be communicated in terms of probability; e.g. 'there's a 60 per cent chance of rain tomorrow', but more often it'll be done in a more qualitative way; e.g. 'We're keeping a close watch on the weather system that's

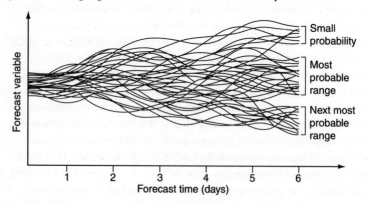

Figure 8.6 Schematic representation of a forecast of a particular variable (e.g. temperature or pressure) from an ensemble prediction system. Each line represents a single ensemble member.

expected to bring rain tomorrow as there's some uncertainty as to how it will develop.' Figure 8.6 shows a schematic representation of this ensemble process at work. A graph of a particular forecast variable (temperature at a particular location, say) is plotted for each ensemble member through the first six days of a weather forecast. By the end of this six-day period, the forecasts seem to fall into three clear groups, and the most probable forecast range is given by the group with the largest number of members.

Post-processing

NWP forecasts are pretty good at predicting the general weather conditions that will affect the region covered by the model, certainly for the next two to three days. The overall pressure patterns and wind directions will be quite accurate, and areas of precipitation associated with frontal systems will be in about the right places although the exact location and intensity may not be spot on. One thing that NWP models aren't so good at is predicting the temperature as measured by a weather station on the ground *at a specific location*. One problem is that the model grid box may cover quite a large area (maybe 100 square kilometres or more) but there is only one temperature at the surface for the whole box. A town within that box may have temperatures that differ considerably from a rural location in the same box. The difference in temperature between these two locations will also be dependent on the actual situation. On a calm, clear night the minimum temperature in the city may be perhaps 5°C (9 F) higher than in the countryside, whereas if the night is cloudy, with a strong wind, the difference is likely to be much smaller. Sometimes the differences in temperature within a grid box may be dependent on the wind direction. If for instance a grid box happens to be on the coast then it will include both coastal locations and places that are perhaps 10–15 kilometres inland. If the wind is blowing from the sea on a winter night then the minimum temperature on the coast may be considerably higher than the inland locations, whereas if the wind is from the land then the two places might well experience very similar temperatures overnight. All of this means that a forecaster can't just take the temperature predicted by an NWP forecast and apply a constant correction to it in order to get a temperature forecast for a specific location. All the complicating factors need to be taken into account, and this takes time. If a forecaster needs to produce overnight minimum temperature forecasts for 100 different towns and cities, they'd need local knowledge for each one, and then would need to take account of the overall conditions prevailing at each place in order to adjust the temperatures predicted by the model. This would take far too long to do manually, or would require a large army of forecasters. Instead, automated computer programs are available to this job in a fraction of the time. These computer programs are referred to collectively as *post-processing*, because they take the output from the NWP forecast and modify it after the model has finished running. There are currently two main post-processing methods in operational use.

Model Output Statistics (MOS)

This method is based on finding statistical relationships between the grid box variables from the NWP forecast and the actual weather conditions observed at specific locations. These statistical relationships need a large database of previous examples to draw on and they may change completely if a new version of a model is introduced. MOS temperature forecasts are generally more accurate than the temperatures taken directly from an NWP forecast, but they can go very wrong. If for instance a particular set of weather conditions occurs which doesn't happen very often, then the database of previous examples won't be much help and so the MOS forecast may be based on almost no previous information. In such a case the human forecaster needs to be on the ball to realize that the forecast provided by MOS is likely to be wrong.

Site Specific Forecast Models (SSFM)

This method is more physically based than MOS, which relies on statistical techniques. A site-specific forecast model is a very detailed model of the area surrounding the place for which it is providing forecasts. The output from the NWP model for the relevant grid box is fed into the SSFM, which then applies the laws of physics again, but this time with much more information about the local conditions. For instance, a SSFM for Heathrow Airport near London would know that the airport is ringed to the south and west by several large reservoirs. Thus in a very light south-westerly wind, on a cold clear night, Heathrow is quite susceptible to fog as the reservoirs act as a source of water vapour to the atmosphere at Heathrow. The NWP model doesn't know that the reservoirs are there and so tends to predict rather drier air at Heathrow than occurs in reality.

Beyond the NWP forecast

Once the team of senior forecasters have assessed the quality of the computer-produced forecast, they can pass their guidance on to the other forecasters who have the job of producing the text and graphical forecasts used by their customers and provided to the general public via the TV, radio and internet. The output from the NWP forecast is a powerful tool and has many uses beyond simply producing pretty graphics for use on TV and the internet. In particular, the output data produced by an NWP model can be used to *drive* other types of forecasting model with very specific applications. Here's a list of some of the potential uses of data from NWP models. Some of these are run routinely within the forecasting agency every time a new NWP forecast is produced. Others are run only in very specific circumstances and some are run by completely different organizations.

▶ **Ocean surface forecasts** – *Ocean waves and swell are essentially driven by the pattern of winds and pressure in the overlying atmosphere. So by plugging the wind and pressure forecasts from an NWP model into a model of the ocean surface, forecasts of wave heights, periods and directions can be made. Shipping and offshore industries such as oil and gas platforms all benefit from these forecasts. Additionally ocean models can be used to predict* storm surges *that occur when winds, low pressure and high tides combine to produce very high waters adjacent to coasts. Much of the damage caused* by land-falling hurricanes is associated with the storm surge rather than the winds or rain.

▶ **Atmospheric pollution forecasts** – *Industrial accidents, sabotage or warfare can lead to major releases of pollutants into the atmosphere. Particularly in the case of nuclear accidents such as the Chernobyl disaster in 1986, we'd like to know where those pollutants are going to end up. Many major meteorological agencies have a pollution dispersion forecasting model that can be run on-demand to predict this. The forecast of wind produced by the NWP model is used to predict where any pollution will end up.*

▶ **Hydrological models** – *River flooding is caused by excessive rainfall, perhaps over a very short period of a few hours for flash flooding in small river catchments, or over longer periods of days or even weeks in the case of large river basins. Melting snow can also lead to river flooding. So rainfall and snow-melt forecasts from NWP forecasts can be used to drive river catchment models predicting flow rates and potential for flooding.*

▶ **Road surface forecasts** – *Highway authorities have responsibility for maintaining safe driving conditions on roads, particularly during freezing weather. Many councils and agencies run site-specific road surface condition forecasting models in order to predict where they may need to treat surfaces for ice or clear snow. The obvious inputs from NWP models are the temperature and precipitation forecasts. These models then perform a detailed heat balance calculation for the road surface, taking account of the heat capacity of the surface material and even the warming effect of traffic passing along the road.*

▶ **Utility demand forecasts** – *Power companies such as gas and electricity providers need to predict demand for their utility. They run their own computer models to do this. The main meteorological inputs into such models are hourly forecasts of temperature, but these models also have to take account of human behaviour which may not be weather related.*

Between the forecasts

Most forecasting centres produce a new NWP forecast every six or 12 hours. After each forecast has been run and the senior team of forecasters have assessed it and issued their guidance to their colleagues, one might imagine that there's nothing left to do until the next run of the NWP model. However, this isn't the case, and forecasters are continually monitoring the weather as it develops, checking it against the forecast, and attempting to predict what might happen in the next few hours. It may be that only a few hours into the period of a forecast, the real atmosphere starts behaving in a way that differs from the most recent run of the NWP model. In such a case the forecasters will need to consider issuing updates to their forecasts. Much more likely is that the broad picture given by the NWP forecast will be largely correct, but the forecasters will be monitoring the details. For instance, the computer forecast may have predicted that heavy showers will affect a large region. But the nature of showers is that some places will be hit by several very heavy downpours while, at the same time, somewhere a few miles down the road will remain completely dry. At the extreme end of the weather scale, a forecast may have been issued with a tornado watch over a large area, but forecasters will be continually monitoring the situation to see if and where any tornadoes are actually starting to form so that more specific warnings can be issued.

The job of monitoring the weather as it develops, and trying to predict what might happen in the next few hours, is called *nowcasting*. It's a rather horrible piece of jargon, but it describes the process rather well and it trips off the tongue somewhat more easily than 'real-time monitoring and very short-range forecasting'. The main tool for this task, and one that is absolutely invaluable to forecasters, is weather radar. Using a network of weather radars, a forecaster can see a picture of exactly where and how heavily rain is falling over a large region, in near real-time, with updates as often as every five minutes. In the case of a band of frontal rain, predicting what will happen in the next few hours is largely based on assuming that the band will move at the same speed as it has done over the last couple of hours, and maintain its intensity. This isn't always a good assumption of course, and frontal rain bands can change their intensity considerably over an hour or so. In the case of showers it is even more difficult, as showers can develop from nothing, rain heavily and die away completely on a two- or three-hour timescale, so just assuming that the current crop of showers will move with the wind without changing much is a very poor assumption. However, it might be clear that the showers are developing in a pattern, perhaps triggered by a particular geographical feature, and this information can be used to predict which locations are likely to experience the heaviest showers. Tools are currently being developed which allow forecasters to spot where showers might be starting to form *before they even appear* on the weather radar. Shower development is always preceded by updraughts – rapidly rising currents that carry air aloft so that it cools and the water vapour in the current starts to condense, forming clouds. These updraughts often carry tiny insects, dust and pollution particles up from near the surface, so if a radar is tuned to identify these particles then the updraughts can be spotted before the clouds even start to form. Sophisticated computer programs are used to process the data from the weather radar network and project it forward in time depending on the prevailing wind direction and large-scale conditions so that the nowcast can be continually refined as new information becomes available. Much of this information never makes it to the general public, and let's face it, most of us don't really need a forecast for the next few hours which is updated every 15 minutes or so. But in cases such as flash flooding or tornado risk, the nowcast forms the basis of the warnings which are issued through TV, radio and internet.

Insight – Nowcasting and sport

Some weather-dependent major sporting events are prepared to pay large sums of money for accurate and timely nowcasting information. The ground staff at outdoor tennis tournaments such as Wimbledon in the UK would like to know exactly when rain is going to hit so that play can continue right up to the last possible minute, but then the courts can be covered before they get soaked. Motor racing teams also like to know if and exactly when rain is likely during a race. Having the correct tyres for a wet or dry track can make all the difference between winning and losing a race, so if you can be warned to get your car into the pits for a tyre change just as the rain is starting then you may be able to steal a march over your competitors. Some Formula One race teams actually employ a meteorologist as part of their pit crew! Golf is another sport with a major sensitivity to weather – in this case thunderstorms. No golfer wants to be standing out in the middle of a golf course, holding a metal club which will act as a very effective lightning conductor, when a thunderstorm sets in. In the US about 70 people are killed annually by lightning strikes and about five per cent of these deaths occur on golf courses. Ten times as many people are injured, sometimes seriously, by lightning strikes. Most major tournaments will take a lightning warning service from a forecasting company or agency to ensure that players aren't put at risk.

Summary

In this chapter we have looked in some detail at how the big weather forecasting agencies go about producing their forecasts. Computerized Numerical Weather Prediction (NWP) models sit at the heart of the operation and are becoming more detailed and more accurate all the time. However, human forecasters with vast experience of the weather are still needed at various stages of the process. They check the information going into the models to create the initial conditions, assess the early period of the forecast to ensure that it is in agreement with reality and make informed judgements about how good the longer-range period of the forecast is likely to be. This allows them to express confidence in the *predictability* of the atmosphere, and recognize the situations when the atmosphere may be inherently unpredictable. This confidence (or lack of it) can be expressed when conveying information to users of the forecasts.

By knowing something about how forecasts are produced and interpreted, we should be able to make better judgements about when and how much to trust the forecast and have a better idea of what NWP forecasts can and can't do. In the next chapter we'll look at how we can use this information, together with local knowledge and experience, to add value to weather forecasts from the media, and also how we can go about producing our own forecasts.

10 THINGS TO REMEMBER

1 *The history of weather forecasting is a fascinating study, involving careful observation of weather systems, application of the laws of physics to the atmosphere and the use of computers to perform increasingly complex calculations to produce increasingly accurate forecasts.*

2 *All weather forecasts today are produced using* Numerical Weather Prediction – NWP.

3 *NWP involves writing as equations all the relevant laws of physics that apply to the atmosphere, and then coding those equations into a complex computer program.*

4 *All NWP forecasts begin with a detailed analysis of the state of the atmosphere at the start of the forecast period.*

5 *Due to the chaotic nature of the atmosphere, small errors in the initial conditions can lead to a rapid growth of errors in the forecast.*

6 *Data assimilation blends many types of observation together with information from previous forecasts to produce the best possible estimate of the state of the atmosphere at the start of the forecast period.*

7 *Senior forecasters make a careful assessment of the forecast as it runs, checking the early stages of the forecast against observed weather conditions to spot any developing discrepancies.*

8 *Ensemble forecasting involves running a forecast many times with slightly differing initial conditions, in order to assess how likely different forecast outcomes are and to spot the possibility of extreme events occurring.*

9 *Output from NWP forecasts is routinely used to drive other types of forecast such as river and coastal flood prediction, road surface conditions, gas and electricity consumer demand and ocean wave and swell predictions for shipping and offshore industry.*

10 *In between runs of the NWP forecast, weather forecasters constantly monitor satellite and radar images and other weather observations to check that the forecast is still on track and to add detail to their forecasts for customers.*

9

Making your own forecasts

In this chapter you will learn:
- *how to expand on weather forecasts from the media by using local knowledge*
- *what use traditional 'weather lore' can be in helping to forecast the weather*
- *how to make weather forecasts by watching the development of clouds and noting changes in the wind.*

There will always be occasions when we want to know what the weather is going to be doing at some point in the future. How much detail we need will depend on what we're intending to do. Sometimes a simple 'rain, yes or no?' will suffice for outdoor activities, building work or farming. We might want accurate overnight temperature forecasts to help us decide whether tender plants in our garden need protection. For coastal sailing we'd quite like to know the wind strength and direction at every point along our route, and for the pilot of a light aircraft, the added dimension of altitude adds yet further complications. While the pilot of the aircraft will have access to very specific weather forecast information which isn't freely available to the general public, for many people, whether pursuing leisure activities or wanting to know how the weather may affect their business, the forecasts available through the media may not be detailed enough. We'd like more specific information. On some occasions, we might not have access to any forecasts at all. Wilderness hiking and camping in a region with no mobile phone coverage puts us beyond the reach of the forecast information we'd normally rely on. In this case, we really do have to be able to make our own forecasts from scratch. Before we start thinking about that, we'll consider ways in which we can add detail to the forecasts available to us through the media.

Interpreting media forecasts

TV and internet forecast providers do a very good job of convincing us that they can give us a detailed forecast for our home, school or workplace. Many internet sites provide a service where you input your postcode or zipcode and obtain a forecast for the next five days, giving the impression that the forecast has been made specifically for your house. Watch many TV weather forecasts these days and you'll see that the cartoon cloud and rain symbols which covered large areas have been replaced by animations of rainfall which show hour-by-hour forecasts of the location of precipitation, with colour schemes that indicate the intensity of the rain. While the quality of weather forecasts has certainly improved over recent years, don't be fooled into thinking that the forecast is going to be spot on for your house. That postcode-based forecast on the internet is probably taken from a forecast from an NWP model with grid boxes at best ten kilometres square. We know that some weather variables, such as minimum temperature, can vary by several degrees over such an area. The animated blobs of rainfall on the TV forecast may give the impression that there'll be a heavy shower over a particular town at 3 p.m., whereas a town 20 miles down the road will remain dry until affected by a light shower at 5 p.m. What that forecast is actually telling us is that there'll be a mixture of light and heavy showers in that general area throughout the afternoon.

So being armed with a basic knowledge of what a computer-generated weather forecast can and can't do is a very useful first step in getting more out of the forecast. The next step is to apply some knowledge of the way local conditions affect the weather wherever we happen to be forecasting for. This is easy to do if we've spent many years observing the weather at a particular location and noting its peculiarities. However, if we go somewhere unfamiliar, for instance on a holiday abroad, we won't have that local knowledge and we'll be far more reliant on the weather forecast as presented through the media. That's not to say that we can't attempt to make our own forecasts even somewhere quite unfamiliar. There are three basic steps to producing a forecast for any given location, which correspond fairly well with Chapters 3, 4 and 5 of this book.

1 *Firstly we need to know whereabouts in the pattern of the global circulation the place we are forecasting for is actually located. This will allow us to make some overall assessment of what the average, or typical conditions for that location are likely to be, i.e. the climate of that location. For instance, if we're going to Jerusalem in July, a glance at the climate record will tell us that it simply doesn't rain in that month, so rain isn't a possibility we even need to consider.*
2 *Secondly, we need to know what specific weather systems will be affecting the location during the period of the forecast. This will allow us to assess whether temperatures will be higher or lower than average, or whether precipitation is likely or not.*
3 *Finally, we then need to think about how the local conditions might modify the effects of those particular systems so that we can produce a location-specific forecast. How much detail we are able to put into that forecast, and how far ahead we can make our forecasts, will depend on the information available to us.*

Forecasting with no aids

In these days of ever-increasing information technology, it is getting harder to imagine a situation where we have no access whatsoever to any weather information. However, it is still possible to be in a situation where you have no access to TV, radio, internet or mobile phone or any meteorological instruments. In such cases, the only available guide is the weather itself. Watching the sky, noting changes in wind strength and direction, or the visibility of distant objects can all give clues to what sort of weather might be on the way. In such a situation, experience and knowledge of the type of conditions that *might* occur is invaluable. Even so, we shouldn't kid ourselves that we might be able to predict the weather more than a few hours in advance. By watching the skies we might be able to predict the development of showers, or detect the approach of a rain-bearing frontal system, but without more evidence we won't be able to say very much about how heavy the rain will be or how long it will last. However, by careful study of the clouds, we should be able to give ourselves enough warning of approaching bad weather to, for instance, get down from an exposed hilltop to a more sheltered location before the bad weather sets in.

WATCHING THE CLOUDS

Let's imagine that we are about to spend a summer day out of doors. A classic summer day may begin with clear blue skies and light winds. As the Sun gets to work heating the ground, small convective cumulus clouds start to develop. Watching these clouds carefully can tell us a lot about how the weather will develop through the next few hours. Here's a checklist to keep in mind when watching the development of cumulus clouds.

1 *The most important characteristic of these clouds is their vertical extent. Even if they become widespread across the sky, if their vertical extent remains smaller than their horizontal dimension then it is unlikely that they will develop into rain-bearing shower clouds. However, if they develop a towering appearance, with sprouting 'cauliflower' tops, then there's a good chance that showers may be on their way. Plate 23 shows a cumulus cloud at this stage of development. No rain is falling from the cloud yet (the base is clearly defined) but the top of the cloud is starting to extend upwards.*
2 *As the clouds develop vertically, keep a close eye on the tops of the clouds. Look for the telltale signs of ice crystals forming.*

*If the cloud tops start to develop a wispy appearance, less
well defined than the hard-edged, lumpy appearance of water
clouds, then there is probably ice in the cloud. The presence
of ice considerably accelerates the growth of cloud droplets,
speeding up and often intensifying the formation of showers.
Of course, the presence of ice in a shower cloud is also
essential to the formation of hail.*

3   *Look for the signs of* organization *in the convective clouds.
Are new convective clouds starting to develop next to clouds
which are already deep and well developed, as shown in
colour plate 15? If this is the case, showers are likely to be
more prolonged than if each convective cloud appears to be
developing in isolation.*

4   *Watch the direction of movement of the clouds. Are they
moving in a direction close to the wind at the surface, or in a
very different direction? If the clouds are moving in a direction
very different to the surface wind, this suggests the presence
of significant vertical windshear (i.e. the wind is changing
direction with height). Windshear can often act to organize
and prolong showers, so this is sometimes a sign that showers
may be quite significant.*

5   *As the clouds become more developed, watch their bases.
A very clear flat base (as in colour plate 23) indicates that
no precipitation is falling yet, whereas a grey, hazy indistinct
base suggests that rain is falling out of the cloud. This is
particularly obvious in cold conditions, when the shower is
falling as snow, which may or may not be melting before it
reaches the ground. Falling snow creates very obvious
'fall streaks' beneath the base of the cloud.*

6   *As clouds develop and move over you it gets harder to see
their development because all you can see is the base of the
cloud blotting out the sky. At this stage be aware of any wind
changes. A sudden, gusty wind springing up may be associated
with a 'gust front' which often precedes the onset of heavy
precipitation by several minutes.*

7   *If the tops of the clouds start to flatten and spread out with
time, without the presence of ice, this is a sign that the
convective development is being suppressed. This in itself could*

mean one of several things. The first is that high pressure may be building or moving into the area. High surface pressure is usually associated with descending air, and this sinking motion in the atmosphere suppresses any convective development. This is a good sign for the weather and may indicate the onset of a settled spell of weather that may even extend for several days.

Another reason for the flattening out of convective clouds is quite the opposite of a developing or approaching region of high pressure. An approaching frontal system, particularly a warm front, may be *preceded* by large-scale descent of air, so the suppression of the convection in this case is a bad sign. Look for the telltale signs of an approaching front, essentially given by the progression of cloud types shown in Figure 4.11 in Chapter 4. Here's a checklist for this situation.

1 High cirrus-type clouds 'invading the sky' from the horizon, are often the first sign of an approaching front. Remember though that most cirrus cloud is not necessarily a bad sign. The wisps of cirrus shown in colour plate 18 for instance may be associated with strong upper-level winds but are probably just isolated patches not heralding any frontal approach.

2 Dense cirrus or cirrostratus clouds starting to fill the sky from a particular horizon often herald an approaching front. Plates 24 and 25 illustrate this process. The cumulus clouds in plate 24 are starting to flatten. The sky directly overhead is still blue, but look lower down on the horizon, and the sky above the cumulus clouds is turning milky-white. In plate 25, taken 15 minutes later, this milky white layer of cirrostratus cloud has spread right across the sky above the cumulus, which is now flattening out into a layer of stratocumulus. Two hours later it was raining heavily from an approaching warm front.

3 Even an increase in the number and persistence of aircraft condensation trails could be a sign of an approaching front. Increasing humidity in the upper troposphere associated with a frontal region helps the formation of these contrails and means that they will generally last longer once formed.

4 *Cirrus and cirrostratus don't by themselves produce rain. Look for further signs of the cloud thickening and turning from white to grey. This usually indicates a transition from high clouds formed of ice to medium-level clouds such as altocumulus and altostratus formed of water droplets. Plate 16 shows a layer of altocumulus preceding a warm front. At this stage the cloud is still thin enough for the Sun to be visible through it. As the cloud thickens and lowers, the Sun will become less clear until it is no longer visible.*

5 *The approach of frontal clouds is often accompanied by a reduction in visibility, particularly once rain starts to fall. Remember that heavy rain consisting of large raindrops actually reduces the visibility less than drizzle consisting of many very small drops, so watching the changes to visibility can give an indication of the nature of the precipitation before it actually reaches us.*

The approach of frontal systems is often signalled first by changes in wind direction. Throughout the UK and western Europe, warm fronts often arrive on a westerly or south-westerly flow, and so the establishment and subsequent strengthening of an airflow in the sector between south and west is often an early warning for warm frontal systems, and at this stage, keeping an eye on the clouds will provide further clues. This wind direction can however cause an additional headache. An airflow arriving on the western fringes of Europe from the Atlantic will be laden with moisture. A little uplift due to the presence of even quite modest hills can cause clouds to form, giving a solid layer of stratus or stratocumulus clouds. These low clouds will then obscure any cloud development happening higher up in the troposphere, making it hard to tell whether a front is approaching or not. This is a very common situation along the hilly western fringes of the UK and on the Pacific coast of North America. In such cases, in the absence of any other information, it is probably best to assume that some precipitation will occur and keep the waterproofs handy. Even if there isn't a front approaching, the combination of the warm, moist airstream and uplift over the hills can often be enough to create continuous drizzle. Frontal cloud running over the top of a layer of lower, stratus cloud is doubly bad news to anyone out in the hills. As the rain falls from the higher frontal cloud through the lower stratus layer, its intensity is enhanced by the seeder-feeder mechanism described in Chapter 5. This is one of the main reasons why rainfall in hilly regions is often considerably heavier than surrounding flatter areas. Bear in mind that this effect is also often not well captured by weather forecasting models, so if you are looking at a forecast produced by a computer model, it is often a good idea to mentally add a fair bit to any predicted rainfall rates over mountainous regions, especially those exposed to a moist airflow from an ocean.

There are many other ways that keeping an eye on the sky can help us forecast the weather, at least in the short term. For instance, for coastal sailors, the appearance of cumulus clouds over land during the morning is often an indication that a sea breeze is setting in. If those clouds appear to move inland, with a clear region near the coast then this is an indication of continued development and possibly strengthening of the sea breeze. Bear in mind that complicated coastal geography can complicate sea breeze patterns to the extent that the sea breeze no longer blows directly towards the land.

USING WEATHER LORE

There are a large number of phrases and sayings related to watching the weather and predicting what is going to happen. Much of this *weather lore* goes back perhaps many centuries. 'Red sky at night, shepherd's delight' (or sailor's delight, depending on where you come from) is a commonly quoted one. It would be wrong to dismiss completely these 'old wives' tales' out of hand. Remember that, in the past, man was generally much more dependent on the weather so needed to have a good 'feel' for what the atmosphere might do next. So these weather lore sayings developed often as the result of careful observation. However, there are three major issues with using these things as guidance for a weather forecast. The first problem is that many of these phrases are very location-specific, and a rule that applies in one particular area may not be applicable somewhere else, even a few tens of miles away. The second problem is that the guidance they give is usually very short range indeed, usually applying to the next few hours. Finally one gets the impression that meteorological accuracy has often been sacrificed to get a phrase that rhymes nicely!

Insight – 'Red sky at night'

The 'red sky at night' example quoted above is one phrase that people still set some store by today. It implies that a red sunset is a good sign for the weather, whereas a red sunrise is bad news. Why might this be? The best guess is that this phrase applies to places such as the UK where the weather is dominated by frontal weather systems arriving from the west. So a red sky at sunset might be due to light from the setting Sun in the *west* reflecting off medium- and high-level cloud trailing a frontal system which has already passed through and is moving away to the east. However, a red sky in the morning might be due to light from the rising Sun in the east, reflecting off the high- and medium-level clouds of a frontal system approaching from the west. So in a region where the weather is not dominated by frontal systems travelling from west to east, this phrase won't apply. Even in the UK not every weather system approaches from the west, and not all high- and medium-level clouds which turn red at sunset or sunrise are associated with fronts.

Levels of atmospheric pollution can also affect the colour of sunsets and sunrises. Colour plate 27 shows a vivid red sunset over the UK but this photograph was taken on 5 November, when smoke from bonfires would have added a lot of small particles to the lower atmosphere, causing increased scattering of the light from the setting Sun, turning the sky red. Large volcanic eruptions also inject many tiny particles into the high atmosphere which can be carried around the globe by strong upper-level winds, leading to periods of dramatic sunsets and sunrises.

There is a lot of 'weather lore' related to the behaviour of plants and animals. Much of this claims to be able to produce forecasts for the coming season, i.e. several months in advance. This type of forecasting is pretty well meaningless. Plants and animals respond to the weather conditions that are happening at the present time, or in the recent past, and it is wrong to credit them with some sort of mystical ability to know what will happen in future. Take the well-known example of 'Oak before Ash, we'll have a splash. Ash before Oak, we'll have a soak.' This is based on the appearance of leaf buds on oak and ash trees during spring. The idea is that if the oak leaves appear first, then the coming summer will be a dry one, with only a 'splash' of rain. If the ash leaves appear first however, the summer will be wet. The emergence of tree leaves depends on a number of different factors such as soil moisture levels, temperature and light levels. Trees draw up water through their roots and then lose that water through their leaves as part of the photosynthesis process. So if the soil is very dry in spring due to a dry winter, then some trees may delay putting out their leaves in order to conserve water within the body of the tree. Different trees have different tolerances to this type of factor, and so this may affect the time at which their leaves appear. For instance, shallow-rooted trees will be more sensitive to a lack of moisture in the soil than deep-rooted ones. Oak trees are generally deeper rooted than ash trees and so are less sensitive to dry soil because they can draw up water from deeper below the ground. Ash trees are more sensitive to soil moisture levels and so late appearance of leaves on ash trees may be a response to a dry winter, but is certainly not a sign that the coming summer will be dry. This particular example is definitely one where poetic considerations have been placed above scientific accuracy!

On a shorter timescale, the behaviour of birds and animals is sometimes used as a weather forecasting aid. On very short timescales, say a few hours, this type of weather lore may be of some use, as birds and animals are perhaps more sensitive to the weather conditions than humans and so will start responding to changes in the weather before we've noticed them. Let's consider another well-known example. 'When swallows fly high, the weather will be dry.' Why do swallows fly in the first place? They fly in order to catch the insects that they feed on, and so the height at which they fly will be a response to how far above the ground these insects are to be found. The insects themselves have very little choice about how high they are flying, as they are carried around in the lowest part of the atmosphere by the wind and vertical air currents. In good weather, when the Sun is heating the ground strongly, vertical currents of convection called *thermals* will develop which can carry the insects many hundreds of metres above the ground. However, in cloudy conditions, these vertical currents will be suppressed and so the insects will be found nearer the ground. However, caution must be applied in this case. The thermals that carry the insects high into the air may be a precursor of the development of showers, and so swallows flying high may sometimes indicate precisely the opposite of dry weather. In this case, watching the sky and the formation of convective clouds will be a much more reliable guide to the weather over the next few hours than the swallows.

The message here is a simple one. Most 'weather lore' was formulated before man understood how the atmosphere works. Now that we understand the systems that bring our weather and how they form, develop and decay, we don't need to rely on rhyming couplets formulated several hundred years ago to help us make our forecasts. However, some of the phrases and traditions are fascinating and it can be good fun trying to understand how they have come about in the first place.

Aids to forecasting

Watching the sky and checking the wind direction will only help us to produce weather forecasts for very short periods – certainly less than 24 hours, and in most cases less than six hours ahead. However, the internet has revolutionized weather forecasting, both for the amateur and professional forecaster. So much meteorological information is available at the click of a mouse these days to anyone who wants it. Even as recently as the mid-1980s, a professional weather forecaster at a weather centre, military airfield or civil airport had access to perhaps ten rather poor-quality satellite images a day. Rainfall radar data was available but it was somewhat unreliable and there were still voids in the network. Nowadays much more detailed information is freely available on the internet, including high-quality satellite images updated every hour, animations of images from rainfall radar networks, observational data from weather stations and even forecasts from many of the world's major computer weather prediction models. Tapping into this huge resource makes the production of one's own weather forecasts a real possibility, and in fact, many small, specialist weather forecasting companies have appeared over recent years due in large part to the availability of meteorological data for free via the internet. At the back of this book the Taking it further section lists some of the best websites to use when looking for meteorological information. This list is by no means exhaustive, and by making your own searches and following links, there are plenty of other sites out there to be discovered.

A forecast process

Having found and become familiar with some of the excellent resources available via the internet, it is a good idea to follow a consistent *forecast process* when trying to make your own predictions. In this way, assessing the available information, and interpreting it in a way that will help you produce predictions will become an almost automatic process.

STEP 1 – GET TO KNOW THE CLIMATE

A prerequisite for making weather forecasts must be to get to know and understand the *climate* of the place you wish to forecast for. What are the normal weather conditions like at different times of the year? What types of weather system affect the area of interest? How does the weather vary depending on wind direction? This is fairly easy if we are forecasting for somewhere we have studied the weather over many years at first hand, but is more difficult if we are forecasting for somewhere unfamiliar. Once again there are websites and books that will help us with this and some of these are listed in Taking it further. Knowing where the location sits in relation to the global circulation discussed in Chapter 3 will also help us understand the climate of that location.

STEP 2 – LOOK AT THE BIG PICTURE

Once we have a good feel for the climate of a particular place then we can start thinking about making forecasts of the actual conditions on any particular day. The best way to start this process is to take a look at the weather conditions over a wide area around our particular place of interest – what meteorologists sometimes refer to as the *synoptic* or broad-scale picture. This means a radius of perhaps several thousand kilometres around the site we are actually forecasting for. By doing this we'll make ourselves aware of the general direction from which the weather will be arriving, and will see any weather systems that might affect us well before they arrive. This ensures that a fast-moving system that we hadn't noticed because we only looked at the very local weather conditions doesn't catch us out.

The best way to form this 'big picture' is to look at a combination of satellite pictures and surface pressure charts. Infrared satellite pictures from the main geostationary satellites, giving coverage of the whole globe (see Chapter 7), are available via the internet on an hourly basis. Learning to interpret the patterns of cloud in these images is an excellent way to get a picture of the locations and strengths of weather systems, and looking at animated sequences of these pictures is perhaps the best way of seeing how these systems are moving. Remember that regions of bright white cloud on an infrared satellite picture indicate deep cloud which may well be associated with precipitation. The shape of the region of cloud will also give an indication of the nature of the weather system. Elongated bands of cloud are usually associated with fronts, whereas large masses of cloud are probably associated with depressions. In the tropics, we can easily recognize the almost circular cloud patterns with spiral cloud bands associated with cyclones or hurricanes, and smaller blobs or lines of deep cloud are often associated with clusters of convective cloud producing very heavy rain. Looking at a single picture will give us a useful snapshot of the broad-scale weather patterns, but it is by looking at a *sequence* of pictures over a period of perhaps six hours or more that we can really start thinking about forecasting. If the clouds associated with a particular weather system are becoming brighter over time in the infrared satellite picture then this indicates that the system is developing with time. Vertical motion is occurring within the system which is deepening the clouds and increasing the likelihood of precipitation. The opposite is probably true of cloud regions that are becoming less bright with time. Changes to the shape of weather systems are also a clue to development. Initially shapeless blobs of cloud which appear to become more organized and concentrated are usually a sign of developing weather systems. The appearance of spiral cloud bands and a distinct eye in a tropical storm are clear indications of the intensification of the storm, and in fact the major storm forecasting centres use the visual appearance in satellite images of tropical cyclones to make estimates of the associated wind strengths.

In a book of this size it just isn't possible to describe in detail all the patterns in satellite pictures that give clues to how the weather will develop. Spotting the signals and recognizing their significance is something that takes even professional meteorologists a considerable length of time to learn. Cloud patterns on satellite pictures also differ considerably depending on location, so that a forecaster working in the US will be looking for different patterns than one working in western Europe, and a tropical forecaster will be looking for yet another set of patterns. By looking at satellite images on a regular basis, you will soon start to recognize some of the most common features and then you can start to form a picture of how these features will affect your weather. Part of the fascination is that every single weather system is different from any other previous system. This means that you need to recognize not only the features that weather systems have in common, but you also need to spot the subtle differences and interpret what these differences might mean for the forecast. If you're serious about learning how to use satellite images then see Taking it further for details of books on the subject.

Once you've looked at the available satellite pictures, the next stage in the process is to associate the features you see with actual weather features. The best way to do this is to look at a surface pressure chart – what meteorologists usually call a *surface analysis*. This is simply a map of the surface pressure over a region, with contour lines called *isobars* joining points of equal pressure, in much the same way that on a geographical map the contours join points of equal height. The positions of any frontal systems are usually added to this analysis by a senior forecaster in the organization that produced the analysis.

Such charts are widely available on the internet for pretty well every part of the globe, and the website of your national meteorological agency is often a good place to find them. Figure 9.1 shows a surface pressure analysis from the Met Office.

These charts show at a glance the positions of any major weather systems – anticyclones, depressions, tropical storms, etc. The direction of the isobars indicates the direction of large-scale airflow. Remember that in the northern hemisphere air near the surface moves parallel to the isobars with lower pressure to the left of the motion. In the southern hemisphere the situation is reversed so that low pressure is to the right of the motion. Here's a checklist of things to think about when studying a surface pressure analysis.

1 *The direction of the airflow affecting your particular location is the first thing to look at for an indication of whether the air will be warm or cold, humid or dry. Think about how the air will be modified as it comes towards you. Air coming from a cold region, and moving over progressively warmer surfaces will become more convectively unstable, and so showers are a possibility. Air from a warm source region moving over cooler surfaces will become more stable, and so showers are unlikely, but if the air is moist from passing over an ocean surface then widespread low cloud or fog is more likely.*

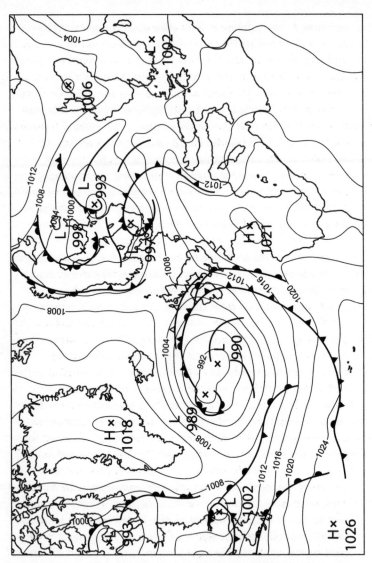

Figure 9.1 Surface analysis chart, showing isobars, fronts and centres of high and low pressure.

2 *What are the main features of the pressure map in your region of interest? In the example of Figure 9.1, the north-western parts of the British Isles are close to a centre of low pressure,* *indicating generally unsettled weather. The frontal systems associated with this low are lying across the west of the UK and will almost certainly be associated with a band of clouds and rain. There is a centre of high pressure over south-west France and northern Spain indicating that the weather there will be largely settled. A ridge of high pressure extends from this centre across most of central Europe, associated with dry, largely cloud-free weather for much of the continent.*

3 *The intensity of individual weather systems is also clear from a surface pressure analysis. Depressions with a particularly low central pressure will tend to have heavy rain and strong winds associated with them. However, bear in mind that this rule tends to break down in some regions. In the Mediterranean for instance, some depressions bring very heavy rain but don't have particularly deep central pressures, particularly in the autumn. This is because the Mediterranean Sea is still very warm at this time of year and that warmth can drive convective activity, enhancing the cloud and rain formation in what otherwise appears to be a fairly weak depression.*

4 *The strength of the winds can be gauged by the spacing of the isobars. Closely spaced isobars indicate a strong pressure gradient, and that in turn indicates very strong winds. Some surface pressure charts even have a scale attached by which the spacing of the isobars can be used to calculate the strength of the geostrophic wind – that is the theoretical strength that the wind would have if friction, and various local effects were not acting to slow the wind down. There are times when simply looking at the spacing of the isobars to gauge wind strength can be misleading. In a small depression with a very low central pressure, the isobars will be packed very tightly indeed. A simple geostrophic wind calculation will indicate winds of perhaps hurricane strength. However, because of the tight curvature of the flow, the actual winds will be considerably less than this calculated value, although still easily up to gale force. The opposite is true around the flanks of an anticyclone. The isobars can be quite widely spaced, but if they are also curved then the actual wind speed can be considerably greater than the geostrophic value.*

Thus while the wind is usually fairly calm near the centre of a region of high pressure, the flanks of an anticyclone can be deceptively windy.

Studying a sequence of surface pressure analyses, together with a corresponding sequence of satellite images, can give a good picture of how the weather systems are moving and developing. This will allow us to see which features are likely to be affecting our own area of interest over the next 24 hours or so, and whether they are likely to significantly influence the weather when they arrive. There are some simple rules of thumb to help us with this.

1 *Fronts tend to move at a speed that is proportional to the geostrophic wind and perpendicular to their orientation. If there are no isobars actually crossing the front then the front will tend to remain fairly stationary.*

2 *The development of a frontal wave on a cold front will tend to slow the progress of the front. Figure 9.2 shows how the development of a wave on a cold front slows down and changes the orientation of the front. Waves can usually be identified first on satellite pictures as the band of cloud associated with the front develops a bulge on the poleward side.*

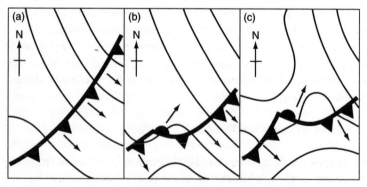

Figure 9.2 Development of a frontal wave, showing fronts, isobars and the direction of frontal movements. An initially straight cold front in (a) is moving from north-west to south-east. A developing wave in (b) slows the progress of the southern end of the front, changing its orientation. In (c) the wave is running along the front, taking the western portion of the front back towards the north-east.

3 *Once a wave on a front has been identified, watch it closely. Most waves indicate a region of enhanced precipitation on a front. However, some waves do not intensify very much, but instead run rapidly along the front, bringing a brief spell of heavier rain. Other waves can deepen considerably and may form small but intense areas of low pressure which can be associated with very strong winds and heavy rain. Infrared satellite pictures are the best tools for monitoring this type of development. If the cloud around the wave appears to be getting brighter (i.e. colder and therefore higher) with time, the wave may well be developing into a depression.*

4 *A front moving into an area occupied by an anticyclone will tend to decay in strength, with cloud dissipating and precipitation dying out. This is because the broad area of descending air associated with the anticyclone will suppress the vertical motion associated with the front.*

These pointers are all associated with the movement of fronts and depressions – the typical weather systems that affect the weather in mid-latitudes particularly during the autumn, winter and spring. Of course, in other parts of the world, and at other times of the year, these features may not be the major influences on the weather conditions. In such cases it may well be smaller-scale features and weather systems that are influencing the weather most and so we need to start looking at more local detail. Even if we are concerned with large-scale features such as fronts and depressions, we need to consider how they will actually affect our own forecast location, and this brings us to the next stage of the forecast process.

STEP 3 – CONSIDER THE LOCAL DETAIL

Having determined where the air that will affect our forecast is coming from, and which particular weather systems will be influencing the weather, we can now start to think about how local factors will affect the forecast. This is where some of the ideas covered in Chapter 5 will start to come into our thinking. Knowledge of local geography becomes an important factor in the forecast process. Even professional weather forecasters find that it takes some time to adjust when moving from one location to another as they get to know how the weather is affected by the geography.

There are two different scales that we need to consider when we start to consider how local conditions will affect our weather. First we must consider how the larger-scale geography will affect our weather and then we can use our knowledge of how the very local conditions will modify the weather further. In some weather situations, the first of these two scales will be the most important, whereas in other cases, the smaller scale will dominate.

On the larger scale, there are two important factors that we must consider. The first is whether we are forecasting for a location on land or at sea, and if the location is on land, does the prevailing wind direction mean that the location will still be influenced by the presence of the ocean? The second important factor is the regional geography. How high above sea level is our location, are there any nearby hills or mountains and if there are, how will they affect the weather where we are?

Remember that wind speed will generally be reduced over land regions compared to the strength of the wind over an open ocean surface, by an amount that depends on the nature of the land surface itself. This reduction can be quite considerable. Over land the wind speed is generally only 50 per cent of what it would be over an ocean surface, and in many inland locations the wind speed will be perhaps only 25 per cent of that over an ocean. Gale-force winds over inland areas are therefore quite rare, although exposed coastal locations will often feel the full force of onshore gales, and in some cases local coastal features such as headlands can even result in a speeding-up of the wind.

Coastal locations with an onshore wind will generally experience higher humidity than inland locations, and this can lead to low cloud formation at the coast as moist air from over the sea is lifted slightly as it passes over land. If however, the air is convectively unstable, clouds are more likely to form inland as heating by the surface as the air moves inland forces air parcels to rise until they form cumulus-type clouds – see the satellite images in plate 9 for an example of this type of situation. Coastal regions tend to have a smaller diurnal temperature range than inland areas. During the day, the ocean surface warms far less than a nearby land surface and so maximum temperatures at the coast tend to be lower. However, at night the ocean surface also cools far less than a land surface, so minimum temperatures near the coast will be higher than inland. The magnitude of this effect is itself dependent on the large-scale conditions. It will be most marked in calm, clear sky conditions and much less apparent during strong winds and cloudy conditions. The contrast between the temperature of the land and sea surfaces can also lead to the formation of a sea breeze – see Chapter 5 for more details.

The effects of altitude are generally to increase wind speed and reduce temperature. These effects can vary considerably depending on both large-scale weather factors and local conditions, so it is hard to give definite rules of thumb. However, when in the mountains you can expect the temperature to cool by about 6–7°C (11–13 F) for every one kilometre (3,300 feet) gain in altitude. This means that a balmy 21°C (70 F) at sea level will translate to freezing temperatures at an altitude of 3,000 metres (about 10,000 feet). The effects of altitude on wind speed are more difficult to quantify than temperature as the shape of the mountain and proximity to other high ground will have a big impact. In the case of very high mountains such as the highest Himalayan peaks, the tops of the mountains may even be high enough to be affected by jet streams, leading to winds of well over 100 mph.

Hills and mountains generally enhance cloud and precipitation amounts. A map of rainfall amounts over the UK for instance looks, with a very few exceptions, very similar to a height contour map – as shown in Figure 9.3.

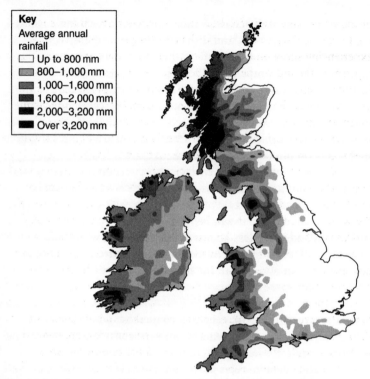

Figure 9.3 Annual average rainfall for the UK and Ireland. This could just as easily be a contour map of the country.

Even low ranges of hills such as the Chilterns in central southern England act to increase the rainfall. The windward side of any mountains will usually experience more rain from a given weather system than the sheltered, leeward side as the seeder-feeder mechanism (see Chapter 5) gets to work. Here again, it is difficult to give exact rules of thumb for how much heavier the rain will be over hills and mountains. Amounts can vary hugely over very short distances due to quirks in the local landscape. If forecasting for a hilly or mountainous area, it is safe to assume that it will be wetter, and sometimes much wetter, than an equivalent site at sea level. Of course, the temperature decrease with height associated with mountains means that a weather system that is producing rain at sea level may actually be bringing *snow* to the hills. This is a very important factor to consider if setting out for a walk or climb in hilly or mountainous regions. Even relatively modest mountains such as the English Lake District, at less than 1,000 metres high, may be experiencing snow on the summits while the nearby Cumbrian coast is getting rain and temperatures of 5°C (41 F). Think about this before setting off on a mountain walk and pack appropriate clothing and equipment. Finally, it should be noted that weather forecasting for mountain areas is a very specialized subject, and we can only skim the surface here. See Taking it further for a list of books written specifically on this subject if you want to learn more.

One tool that is invaluable when trying to assess local effects on rainfall is meteorological radar (see Chapter 7). In many parts of the world, extensive networks of these radars provide detailed, real-time information on the distribution of precipitation. Images from these radars are often available via the internet, so if we can get our hands on the picture for our own forecast area, we can really see how rainfall amounts are varying over the area and how quickly the rain is moving. Internet compatible mobile phones mean that we can even access these pictures while outdoors, in the hills or even at sea in coastal waters. Many national weather services provide the latest rainfall radar information on their websites, and some even provide animations of the pictures over the last few hours. This is a particularly useful tool as it allows us to assess how fast an area of rain is moving. We can then use this information to predict how the rain will move over the next few hours. This works best on frontal rain bands and less well on showers which are developing and decaying all the time. Be aware that bands of frontal rain can speed up or slow down through time, and can also strengthen and decay, so simply assuming that they will move at a constant speed and maintain their intensity is slightly risky. Rain will generally intensify as it passes over hills and mountains, as discussed above. Particularly in the summer, an apparently weak band of rain can sometimes intensify rapidly during the day as surface heating triggers convective activity along the front.

Once the larger considerations of terrain have been taken into account, we can start to think about the much more local factors that affect the weather. This is where we might be able to improve on a weather forecast produced by a large weather forecasting agency since we may have a better idea of our own very local weather conditions. Since every location is to some extent unique it is hard to give hard and fast forecasting rules when we get down to this scale, as local knowledge and experience become important factors. The sort of things we may be considering here are the effect on temperatures of urban areas or rural frost hollows, very local wind patterns due to hills or coastal features, the effects on fog formation of local terrain and proximity to bodies of water or the generation of showers by nearby hills or coasts. Chapter 5 goes into more detail of how these sorts of factors will affect the weather, but there is no substitute for local knowledge. If trying to forecast for somewhere new, talk to the locals to find out about any local peculiarities in the weather. Often farmers and people who spend much of their time outdoors are the best people to provide this kind of detail. But if you can find a local meteorologist, either a professional forecaster or an amateur enthusiast with their own weather records, then so much the better as they may be able to help you quantify the local effects rather than simply describing them.

Using forecast information from the internet

As well as the satellite pictures, surface pressure charts and rainfall radar pictures which are available on the internet, there is also a vast amount of *forecast* information available at the click of a mouse. Much of this has been processed to give forecasts for specific locations or particular activities. However, it is also possible to find plots of output from NWP computer models that you can then use to make your own weather forecasts. Much of this information is the stuff that the professional forecasters look at on a daily basis. This gives you a chance to see the information that influenced the professionals to issue a particular forecast, and it also allows you to pit your wits against them and see whether you'd issue a different forecast based on the same information. Bear in mind that nowhere near all the information that the professionals get to see makes it onto the internet, and also that the professionals benefit from constant discussion with their colleagues. They also have the backing of teams of experts who are constantly assessing the performance of the computer-based forecasts and feeding information to the forecasters to let them know what situations the models handle well and whether they have any systematic biases. Add to this the benefit of professional training courses and constant updates on the latest forecasting techniques and it is unlikely that you'll be able to beat the professionals at their own game. It's fun to try though, and studying the NWP forecasts on a regular basis, alongside the available satellite pictures, pressure charts, rainfall radar and other weather observations, is one of the best ways to develop your knowledge of how the atmosphere works and why we get the weather that we do. Taking it further lists a number of websites which provide output from NWP computer models. There are several steps to understanding and using this information.

1 *Look at where the information has come from. Which weather forecasting service provided the data? Did it come from a global model or a regional or local scale model? Check also the time of the information. Is it from the most recent forecast, or from an older forecast that may now be out of date? And what time is the forecast for? Remember that the accuracy is likely to decrease with forecast range.*

2 *Study any key or legend, and make sure you understand what quantity the chart is showing you. For instance, a chart showing a forecast of temperature may not be for temperature at the surface but may be for some other level in the atmosphere.*

3 *Try to relate features in the forecast to features that you can see in the current weather conditions. Can you track a particular band of rain or depression from its current position through the period of the forecast? Bear in mind that at longer forecast times, new weather features may have formed which were not present at the start of the forecast. Identify the weather features that will influence the weather at the location for which you are forecasting and watch how these features develop with time.*

4 *Remember that forecasters at the major weather agencies rarely use the output from just one NWP model to produce their forecasts. Instead they look at the output from several such models and come up with what they consider to be the 'best consensus forecast'. While this takes a great deal of skill and experience to do well, you can try to do the same thing if you can get hold of forecasts from more than one NWP model. If you have data from say four different NWP models, then if three of them are saying that it will be dry in 48 hours time, but one of them is forecasting rain then the balance of probabilities is for dry weather. However, if the one model that is forecasting rain happens to be the best performing model for your particular location, then it might be safer to give more weight to that particular model and forecast wet weather. How do you know which model is best for your location? In short, it is virtually impossible to know this without a great deal of experience or the backing of a large weather forecasting agency. However, there are several online discussion boards where amateur and professional weather forecasters exchange views on the forecasts, and by tapping into these you may start to get a feel for this. These discussions do tend to be liberally sprinkled with jargon though, so don't expect to make much sense of them at first. The other thing to remember about online discussions is that, although someone may sound like they know what they're talking about, they may know less about the subject than it would appear. Just because someone has an interest in the weather doesn't mean they really understand how it works. Filtering out the junk from the good stuff is another skill that takes time to develop.*

Summary

There are many ways that we can add value to existing weather forecasts, and even produce our own. The wealth of resources available through the internet means that we can see some of the information that professional forecasters use when making their own forecasts. Remember though that the professionals have more information than we can get hold of, have been through an extensive training programme and often have many years of experience. This means that the best basis for any weather forecast is the information provided by the large forecasting agencies. Where we can improve on these forecasts is by considering geographical details that may modify the weather locally. We can also monitor the weather as it develops, by watching the skies or by using satellite and rainfall radar data in real-time from the internet. This will allow us to see if the weather is evolving as predicted or is starting to diverge from the forecast. Experience counts for a lot when trying to produce weather forecasts though and you shouldn't expect to be able to beat the professionals at their own game on a regular basis. However, there's a great deal of satisfaction to be gained from making your own forecasts, even if for only a few hours ahead, and then seeing them turn out to be pretty well correct.

10 THINGS TO REMEMBER

1 *Making weather forecasts requires a great deal of skill and experience, and it is unlikely that you'll be able to do better than a professional forecasting agency. It's still fun to try though, and very satisfying when your forecast proves correct.*

2 *Understanding the limitations of forecasts presented in the media can help you interpret them and use them more intelligently.*

3 *In order to produce your own forecasts it is important to understand the climate of your region of interest and know about the different types of weather system that are likely to affect the area.*

4 *Get to know how the local geography affects your weather. Hills, coastlines, large bodies of water and even the vegetation can all affect the weather, and knowing about these things is the one way in which you may be able to produce more accurate forecasts than a large forecasting company or agency.*

5 *When trying to forecast the weather in an unfamiliar region, make an effort to understand the climate of the region and talk to the locals about any particular local weather conditions which may be peculiar to the area.*

6 *Trying to predict the weather by watching the clouds can, with practice, allow you to forecast the development of the weather several hours ahead but is unlikely to yield much information at longer forecast ranges.*

7 *Weather lore – old sayings about how the weather will develop – may also have some limited use in producing very short-range weather forecasts. However, at longer time ranges most weather lore is pretty useless. Remember that birds, animals and plants respond to the weather they are experiencing at the time and do not have mysterious forecasting abilities.*

8   *The internet is a massive resource for amateur weather forecasters, providing links to up-to-date satellite and radar images and all kinds of other weather observations.*

9   *When producing your own forecasts, always work through a set forecast process, looking first at the large-scale weather systems affecting your region and then focusing in to more local detail which may be affected by local geographical conditions.*

10 *Take care when using data from NWP forecasts that you find on the internet. Be sure you know what you're looking at, how up-to-date it is and how relevant it is to the forecasts you are trying to make.*

10

Climate change

In this chapter you will learn:

- *what the greenhouse effect is*
- *how recent weather records indicate that the climate is changing, and how this can be explained by man-made emissions of greenhouse gases*
- *how the climate is predicted to change in the future*
- *what can be done to reduce the impact of climate change, and to adapt to the unavoidable aspects.*

Climate change is big news. Any media story about the weather nowadays comes with a climate change angle attached. Was global warming to blame for this drought, that flood or any particularly damaging or extreme event? Yet it seems not long ago that climate scientists were telling us that the Earth was on its way into another ice age. Now it seems that the Earth is warming up at an unprecedented rate, but just when we've come to terms with that message, we're told that one of the possible consequences of the warming may be that the climate of certain parts of the world may end up getting suddenly colder. All in all it is hard to get the story straight from the way it is presented through the media. Another problem is that one can't help feeling there's a fair amount of vested interest at work on both sides of the story. On the one hand, climate scientists and environmentalists could be seen as pushing the global warming message to ensure continued funding for their work. On the other hand, oil and coal producers, industry and even governments could be seen to be playing down the risks of increased greenhouse gas emissions in order to carry on their businesses with minimum cost and disruption. In this chapter we'll try to separate the facts from the speculation. We'll take a look at the current state of the scientific evidence for climate change and the latest predictions for the future of the Earth's weather and climate.

The greenhouse effect

As discussed in Chapter 2, the greenhouse effect is a naturally occurring phenomenon. Certain gases in the atmosphere such as carbon dioxide, water vapour, ozone, etc. absorb some of the radiation emitted by the Earth's surface. This effectively traps some of the energy in the atmosphere that would otherwise make its way out to space. A natural consequence of this effect is that the Earth's surface is about 30°C (54 F) warmer than it would be if the Earth had no atmosphere. Scientists have also known for many years that an increase in the amounts of the so-called 'greenhouse gases' in the atmosphere would lead to a warming of the atmosphere and the Earth's surface. In fact the Swedish scientist **Svante Arrhenius** published a paper to this effect in 1895. Further research in the 1930s reinforced this message, but at the time the potential warming of the planet was largely seen as a good thing.

Since the late 1950s extremely accurate measurements of atmospheric levels of carbon dioxide have been made at various places around the globe, and they all show a regular, year-on-year, rise in concentrations. The most famous set of measurements is that made at the mountaintop observatory of Mauna Loa in Hawaii (see Figure 10.1). This location is many thousands of miles away from any industrial sources of carbon dioxide, and so the measurements here confirm that increasing levels of the gas are well mixed throughout the atmosphere, and are not simply caused by measurements being taken close to pollution sources.

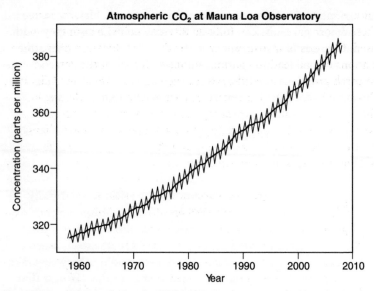

Figure 10.1 Graph of atmospheric carbon dioxide concentrations measured at Mauna Loa observatory in Hawaii.

Levels of other greenhouse gases such as methane are now also monitored closely. Methane for instance is a very potent greenhouse gas with a stronger warming potential, molecule for molecule than carbon dioxide, but with much lower overall levels. Best estimates suggest that global methane levels have about doubled since the start of the twentieth century, with increasing rice paddy cultivation being partly to blame. The other main agricultural source of methane is the huge increase in cattle production. The digestive system of cows produces large amounts of this gas and it doesn't take a lot of imagination to work out how it finds its way into the atmosphere.

The recent temperature record

An accurate picture of global temperature can be put together for the period since the start of the twentieth century, when operational weather stations, with carefully calibrated instruments taking regular measurements, provide us with reliable data. Even these records must be carefully quality-controlled before they are used. Changes in instrument types or the exact location of weather stations can all lead to spurious jumps or trends in the data. Several research groups around the world, such as the team at the Climate Research Unit at the University of East Anglia in the UK, spend their time meticulously checking the records and assembling the best possible unbiased estimate of how temperatures have changed over the last century or so. The graph they have produced, shown in Figure 10.2, is an interesting one.

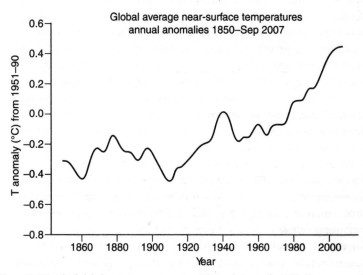

Figure 10.2 Graph of global mean temperature since 1850, relative to the 1961–90 average.

Global temperatures have increased by about 0.8°C (1.4 F), since the start of the twentieth century, but unlike the increases in carbon dioxide, the rise has not been consistent through time. The early part of the twentieth century does show a gradual rise in temperature from about 1910 to 1940. Thereafter the temperature either dropped slightly or remained fairly steady until the mid-1970s when it began to rise again. It has continued to rise since then, and 11 of the last 12 years (1995–2006) are among the 12 warmest years on record.

EXPLAINING THE TREND

The fact that the somewhat erratic rise in global temperature over the last 100 years or so does not closely mirror the steady rise in atmospheric greenhouse gas concentrations might be seen as an argument that increased levels of greenhouse gases are not to blame, and that the observed variations in temperature are natural and random. There are two compelling arguments against this line of reasoning.

The first is that the observed trend in temperature over the past century is almost certainly far larger than any other naturally occurring trend in the past 1,000 years, and so would be hard to explain away by natural variations in temperature. There have been other warm and cold periods, such as the Medieval Warm Period in the twelfth and thirteenth centuries, and the 'Little Ice Age' in the seventeenth century, but the transitions between warmer and cooler periods in the past have been more gradual, and these warmer and colder periods appear to have been more localized, rather than global in scale.

The second argument is that there are good physical reasons why the temperature increase has not been uniform. There are several other, quantifiable effects at work. The first is variations in the amount of energy coming from the Sun, which is the heat source for the Earth and its atmosphere. Solar output varies with the sunspot cycle, and so the amount of energy input to the atmosphere also varies. The second is the effect of volcanic eruptions. Major eruptions, such as El Chicon in Mexico in 1982 and Mount Pinatubo in the Philippines in 1994, inject vast amounts of tiny particles called *aerosols* high into the troposphere and even into the stratosphere. These can remain in the atmosphere for several years after the eruption and have the effect of reflecting incoming sunlight back out to space. So after major volcanic eruptions, a cooling of the atmosphere is usually observed. In fact the Pinatubo eruption was estimated to have cooled global mean temperature by about half a degree over the following year. When these additional effects are included in the interpretation of the global warming graph, the trends can be explained pretty well. Climate prediction centres such as the Hadley Centre at the Met Office in the UK actually run computer models of the climate over the twentieth century in order to see if they can reproduce the observed trends. When only greenhouse gases are included in the models, the fit between the model prediction and the observed trends is rather poor. However, when volcanic eruptions and solar variability are included, the observed trends can be more closely reproduced.

There are also other natural reasons why the temperature trend over the last 100 years or so has not been uniform. For instance, natural fluctuations in global temperature are associated with El Niño. The Earth's oceans can act as a heat store, especially in the tropics. During an El Niño event, when the ocean surface warms across the equatorial Pacific, the ocean passes vast amounts of heat to the atmosphere and so the global mean temperature often peaks in El Niño years. Because El Niño events occur erratically, between every three and seven years or so, this superimposes fluctuations in temperature on top of the global warming trend. The warmest year on record up to the time of writing was 1998 – a year with a very strong El Niño event.

Other evidence for climate change

The instrumental temperature record is not the only evidence that the Earth's climate is changing. There are several other indicators in nature that changes are afoot.

Widespread melting of snow and ice is another sign of increasing temperatures, particularly at higher latitudes where the warming is known to be greatest. Mountain glaciers are generally receding all over the Earth, and permanent ice sheets such as that covering the bulk of Greenland are also in recession. Arctic sea ice also appears to be thinning and receding, perhaps rather rapidly, but there is no evidence that the Antarctic ice cap is melting, apart from in a few locations. Melting ice is contributing to sea-level rise, which appears to be accelerating since the beginning of the twentieth century. However, the largest contribution to global sea-level rise is simply the expansion of seawater as it warms up. Between 1961 and 2003, global sea level rose at an average rate of 1.8 millimetres per year. This may not sound like much, but over the twentieth century as a whole, the best estimate for total sea level rise is about 17 centimetres (6.8 inches). Again, this may not sound like much, but for small island states such as the Maldives and low-lying land regions such as Bangladesh, rises of this order of magnitude can mean a significantly increased risk of flooding at particularly high tides or during storm surges.

Changes in weather patterns all over the globe have been observed which are consistent with global warming. More intense and widespread droughts have been occurring since the 1970s, particularly through the tropics and subtropics. The number of intense tropical storms seems to be increasing in some areas, and over land areas the frequency of extremely heavy precipitation events has increased. Heatwaves are becoming more frequent, and the occurrence of extreme cold temperatures is decreasing.

HOW SURE ARE WE?

There seems to be plenty of evidence of various forms which points towards an upward trend in global temperature which is greater than would be expected from any natural climate variability. But how sure are we that the Earth's temperature is indeed increasing at an unnaturally fast rate, and how can we be sure that this warming is induced by human activity and in particular the burning of carbon-based fuels?

Since 1988, a body known as the *Intergovernmental Panel on Climate Change (IPCC)* has produced regular reports on the state of climate change research, both in terms of assessment of the observed trends and predictions for the future. This body, which consists of an association of some of the world's most eminent meteorologists and climate scientists, brings together all the published research on climate change and, based on all the available evidence, makes assessments of how the climate is changing and how likely it is that any changes are human-induced. The IPCC also makes projections of how the climate will change in the future, again based on as wide a range of evidence as possible.

In 2007 the IPCC issued its Fourth Assessment Report. Since the first report, confidence has increased that human activities are having a warming effect on the climate. The current assessment is that warming of the climate system is *unequivocal* and there is at least a 90 per cent probability that most of the warming seen since the mid-twentieth century is due to human-induced increases in greenhouse gases.

There are still some uncertainties to be addressed however. One major uncertainty is the effect of increased sulphate aerosol concentrations in the atmosphere, also due to industrial processes. It is likely that the observed global warming trend would have been larger if these sulphate aerosols had not been present as they are known to reflect a portion of the Sun's radiation back out to space. They also have an effect on cloud formation. Sulphate aerosol particles act as locations onto which water vapour in the atmosphere can condense as clouds start to form. More aerosol particles mean that the same amount of water vapour will be divided between more, smaller cloud droplets. These smaller droplets are more effective at reflecting the solar radiation back out to space, leading to a cooling effect on the atmosphere. More aerosol may also mean more clouds, but this is far from certain. However, the main message is that sulphate aerosol pollution acts to cool the atmosphere, but the exact amount of cooling is hard to quantify. Sulphate aerosol has quite a short lifetime in the atmosphere compared with gases such as carbon dioxide. If the industrial processes that produce these aerosols are cleaned up, then the amount of aerosol in the atmosphere would decrease very quickly, and this in turn will lead to an acceleration of the warming trend. Carbon dioxide on the other hand remains in the atmosphere for decades and so even drastic cuts in its emissions would have little immediate impact on the climate.

What does the future hold?

As well as assessing the evidence for the global warming that has already occurred, the IPCC also looks at projections of future climate change. The major uncertainty in making such projections is exactly how much carbon dioxide and other greenhouse gases continue to be emitted into the atmosphere. This is extremely difficult to predict. It depends on many different factors such as population growth; rates of industrial development across the globe; willingness of governments and industry to restrict greenhouse gas emissions and move towards cleaner forms of energy generation; and the willingness of individuals to cut down their personal carbon use. This is obviously an area in which climate scientists have no expertise, so economists, engineers and development experts are called in to produce a range of different *emission scenarios* based on different predictions for how the global economy will develop over the next 100 years or so.

These different scenarios can then be input into climate models to produce forecasts of how the climate will change. Climate models are essentially similar to the computer models used to make short-term weather forecasts, described in Chapter 8. There are some important differences however. Firstly, climate models include a detailed representation of the Earth's oceans, since the oceans are a crucial part of the climate system – storing, moving and releasing heat on various timescales. Some climate models also include interactive vegetation, which can react to changes in climate by either dying back if conditions become too hot or too dry, or by becoming more abundant if the climate becomes more conducive to growth. Changes in vegetation can then feed back onto the climate in various ways. One important role of vegetation is to take carbon dioxide out of the atmosphere, but widespread die-back of vegetation would result in a release of this carbon back into the atmosphere, amplifying the greenhouse effect further.

A large number of different climate models are used to produce the projections of future climate change under the different scenarios.

This allows the IPCC to quote likely ranges on their predictions of global warming because some models predict more warming than others. Predictions are made for the amount of warming by the end of the twenty-first century, relative to the 1980–99 average. For a scenario in which energy production moves towards cleaner technologies and industrial development is limited, the predicted increase in global mean temperature by the end of the twenty-first century is between 1.1 and 2.9°C (2 to 5.5 F) with a most likely value of 1.8°C (3.2 F). For a scenario in which energy production remains largely based on fossil fuel burning, the predicted range of global warming is between 2.4 and 6.4°C (5.3 to 11.5 F), with a most likely value of 4°C (7.2 F). These are of course global mean figures, and in some locations, the predicted warming will be much higher than these numbers. Warming is expected to be greatest over northern hemisphere, high-latitude land regions, and smallest over the Southern Ocean and the North Atlantic, south of Greenland. In general, warming will be greater over land than over the oceans. The warming of high-latitude land regions, which are currently covered in snow for much of the year, will accelerate the rate at which snow and ice cover retreats. Plate 26 shows a set of projections of temperature change for different points during the twenty-first century, for a range of different emissions scenarios.

Models are also used to make predictions of sea-level rise over the next century. Once again, these projections are scenario dependent. In a low-emission scenario, the predicted range is between 18 and 38 centimetres (7–15 inches) and in a more fossil-fuel intensive scenario the prediction is for between 26 and 59 centimetres (10–23 inches) of sea-level rise. These values have pretty serious implications for low-lying coastal regions around the world.

Predictions of changes to global mean temperature don't mean that much to most people, who are most interested in how climate change will affect the weather where they live. Predictions of regional or local changes to weather patterns are harder to make than the global trend because there is more disagreement between different climate models at these scales. Even so, there are some patterns emerging that seem to be fairly consistent between the models. Certainly the pattern of warming across the globe, with the largest increases in temperature being over the high-latitude northern hemisphere seems fairly certain. The mid-latitude storm tracks in both hemispheres are expected to shift polewards, with associated changes to wind and precipitation patterns. In northern-hemisphere winter, this movement of the storm tracks will lead to more rain at most locations north of about 50 degrees north, whereas subtropical locations such as the Mediterranean basin will become drier. In northern hemisphere summer the drying around the Mediterranean will be even more marked, with projected decreases in rainfall of greater than 20 per cent in many locations in some of the future scenarios. For the UK the winters are expected to become somewhat wetter, whereas the summers could be significantly drier than the 1980–99 average, especially over the south-east which is already the driest part of the country. Over the US and Canada the picture is less clear-cut. Much of Canada is expected to be wetter in winter, but there is currently much less of a significant signal during the summer. In the southern hemisphere, the trend is also that subtropical regions will become drier, whereas the higher latitudes will become wetter. The exact amount and locations of changes to precipitation are also dependent on the emissions scenario.

In the tropics, the main signal in rainfall is that the tropical ocean regions will become wetter, particularly in the Pacific basin. It is likely that tropical storms will become more intense with stronger winds and heavier precipitation. Still uncertain is how the *number* of tropical storms will change over the next century.

There is also a high degree of confidence that the occurrence of many extreme events will increase throughout the twenty-first century. It is virtually certain that maximum temperatures over most land areas will increase, with an increase in the frequency of heatwaves, such as that experienced by much of Europe during the summer of 2003, being very likely. It is also very likely that the frequency and intensity of heavy rainfall events will increase, with associated increases in the risk of flooding. Despite this, it is also likely that the area affected by droughts will increase, a trend that has already been observed in many regions since the 1970s.

One of the most important things to remember about climate change is that, as well as the predicted, human-induced changes, *natural* climate variability will still be happening all the time. This means that even if the climate is forecast to be warmer in future, periods of cold weather which occur naturally aren't going to stop happening. The further we go into the future, the more likely it is that such colder spells will be less extreme, but a cold summer, or even a run of several cold years in a row is not necessarily proof that global warming isn't happening.

What can be done?

Climate change is a global problem, and as such it demands a global solution. A global reduction in greenhouse gas emissions could be enough to stem the tide of climate change, but these reductions would have to be significant. Due to the nature of carbon dioxide and many of the other greenhouse gases, emissions created today will remain in the atmosphere for many decades, continuing to contribute to the warming trend. At this point, the issue becomes political rather than scientific. Only concerted action by governments around the world can be enough to produce the levels of reductions in greenhouse gas emissions that are considered necessary.

The United Nations has taken the lead in trying to bring world governments together to reach agreement on cutting down on carbon dioxide and other greenhouse gas emissions. In 1992, a summit was held in Rio de Janeiro at which it was agreed that, by the year 2000, emissions of carbon dioxide needed to be stabilized at their 1990 levels. In 1997, at a meeting in Kyoto, Japan, targets were agreed for reductions in carbon dioxide emissions by 2012 for every country. However, some countries were very slow to sign up to agree to meet these targets, and other countries, including the USA, which is the largest single emitter of carbon dioxide, have never signed up to this agreement at all. Countries who have agreed to reduce their emissions are not doing as well as hoped, and in the period 1995–2005, only a very few of the major carbon dioxide-producing countries, such as Germany, have actually managed to *reduce* their emissions. In the developing world, some countries have massively increased their carbon dioxide emissions over the same period, with China showing an 87 per cent increase, bringing their total carbon dioxide emissions up to almost the same levels as the USA.

Further United Nations summits have attempted to get further agreements on reductions in carbon dioxide emissions that will go beyond the 2012 deadline agreed in Kyoto. However, results have been far from encouraging. There are several major sticking points. One is that developing nations see it as somewhat unfair that just as they are starting to become industrialized and their economies are starting to take off, they should be asked to rein in their development and spend money on reducing emissions and switching to other sources of energy. In the developed world, some countries see carbon dioxide reduction as contrary to the best economic interests of their citizens and are reluctant to surrender their economic advantages. All in all, the measures which are needed to curb the threats of climate change will only be taken if there is a major international agreement by all countries and in particular those that are responsible for the majority of carbon dioxide emissions. Such an agreement would need to include concrete targets for emission reductions over the next 20 years or so. In the absence of such an agreement it seems inevitable that further climate change will be a reality with which the world as a whole will have to cope. While some aspects of climate change may not be life threatening, issues such as the increased incidence of heatwaves and droughts in already water and heat stressed regions, more intense tropical storms, sea-level rise and increased risks of flooding are all problems which will probably cost much more to deal with than the cost of reducing carbon dioxide emissions to a level which would prevent these problems.

MITIGATION OR ADAPTATION?

Dealing with the problem of climate change involves two major strands. The first is *mitigation*, where attempts are made to stop climate change happening, or at least to reduce its effects. This largely involves the type of international policy approaches described above, reducing greenhouse gas emissions or looking for processes which might actually remove greenhouse gases from the atmosphere. The second strand is *adaptation* to climate change. This acknowledges that some degree of climate change is inevitable and so we need to plan for how we are going to adapt to this change. The possibilities for adaptation include a massive range of potential actions at all levels, from international policies through to action by individuals. Here are just a few examples.

▶ **Water resource planning** – *In some regions of the globe, some seasons are forecast to become wetter, whereas other seasons will become drier. This implies that levels of water storage need to be assessed so that rain during the wetter periods can be stored to be used during the drier periods. This involves planning at national level, and on very long timescales. Planning for and building reservoirs takes many years, so if we want to be ready for changes to rainfall in the mid-twenty-first century, we need to start planning now.*

▶ **Agricultural adaptation** – *Changes in temperatures and rainfall may mean that crop varieties which are currently being grown may no longer be viable in future climates. Given current projections of climate change there is a lot of research going on worldwide to look for alternative crop varieties that may be able to cope with increased incidence of heat stress or drought. This is a very complex problem. Increased amounts of carbon dioxide in the atmosphere actually mean that most plants can grow more quickly which may be a good thing. However, if a crop matures too quickly it may do so with a reduced yield. Some crop types also benefit from warmer temperatures, but a short period of very high temperatures at a critical stage of development can actually kill the plants or severely restrict their yield. This type of research and planning is particularly crucial for less developed parts of the world where people are far more dependent on local agriculture for survival, and where it may also be harder to persuade farmers to move away from traditional practices.*

▶ **Changes to building practices** – *In the hot parts of the globe, buildings have always been designed with the aim of keeping the occupants cool during the hottest weather. Traditional building styles in these regions include thick walls, small windows and design which encourages airflow through the buildings. In the more temperate regions, this has been less of a consideration, but as temperatures rise, particularly in the summer months, this starts to become an issue. Buildings being constructed now will have lifespans of well over a century, so need to be designed with projections of climate change in mind. This doesn't just mean adding air-conditioning units as this is a very energy-intensive way to keep buildings cool and will itself contribute to the global warming problem. What is needed is a more comprehensive rethink of how buildings are designed.*

▶ **Health planning** – *During heatwaves, numbers of hospital admissions and deaths due to heat-related problems always increase. Planning of hospital capacity needs to take account of the probability that dangerous heatwaves, which currently happen perhaps only once every ten years or so, may become the norm in some countries. For instance, the 2003 heatwave in central Europe resulted in approximately 10,000 extra deaths in France alone due to the hot conditions. However, by the middle of the twenty-first century, the temperatures that occurred in 2003 will actually be thought of as normal, and so plans need to be drawn up now which consider how health services are going to cope with the increase in heat-related illness.*

Summary

There is now little doubt that the climate of the Earth is changing in a way which is outside the range of natural climate variability. Projections of future climate change vary depending on predictions of how much effort the human population makes to reduce its impact, but indications are that we can expect a global warming of *at least* 1°C by the end of the twenty-first century. Regional patterns of temperature and precipitation are harder to be certain about. However, it does seem clear that in many parts of the globe, there will be changes in climate that will require considerable planning and adaptation. In particular the increased risks of heatwaves and droughts present a real challenge to societies, and how we plan for and deal with these risks will become an increasingly important consideration over the coming years.

10 THINGS TO REMEMBER

1 *The natural greenhouse effect, caused by the presence of carbon dioxide and other greenhouse gases, is an essential element that maintains the Earth's climate in a state that is warm enough to support life.*

2 *Increasing levels of greenhouse gases in the atmosphere will result in increasing temperatures as more energy is trapped within the atmosphere instead of escaping to space.*

3 *The average temperature of the Earth shows a lot of variability from year to year compared to the rather consistent increase in greenhouse gas concentrations. This is due to natural climate variability which is occurring all the time.*

4 *Natural climate variability is due to many different factors. Changes in the energy coming from the Sun, volcanic eruptions and uptake or release of heat by the world's oceans can all cause the Earth's temperature to vary on a wide range of different timescales.*

5 *As well as increasing temperatures from direct observations, there is also a lot of other evidence for climate change. Receding glaciers, changes to vegetation and rising sea levels all point to changes in the climate of the Earth.*

6 *Recent assessments of scientific evidence suggest that it is very likely that rises in global average temperatures over the past 100 years or so have been cause by man-made emissions of greenhouse gases into the atmosphere.*

7 *Climate scientists use extremely complex models of the Earth's atmosphere, oceans and land surface in order to predict how climate may change in future.*

8 *There are many uncertainties in predicting future climate change, such as how cloud amounts will change, how the presence of sulphate aerosols will affect the climate and exactly how much greenhouse gas concentrations will increase in future.*

9 *The various uncertainties mean that climate predictions are expressed in terms of a range of possible values rather than as a single number.*

10 *Global warming is a global problem and will only be solved through international agreement and co-operation.*

Taking it further

Recommended books

GENERAL WEATHER AND CLIMATE

▶ *Henson, R. (2007)* Rough Guide to the Weather, *Rough Guides. Published as part of the Rough Guides travel series, this book contains a comprehensive set of climate charts for over 200 holiday destinations around the world. There are also very readable sections on what makes the weather, weather forecasts, extreme weather and climate change.*
▶ *Reynolds, R. (2000)* Philips Guide to the Weather, *Philips. A colourful introduction to weather and climate, packed with weather facts.*

MORE ADVANCED SCIENTIFIC BOOKS ON WEATHER

▶ *Ahrens, C. D. (2000)* Meteorology Today, *Sixth Edition, Brookes/Cole. A great book if you want to get deeper into the science of meteorology. Aimed at university science undergraduates, this book is lavishly illustrated and is accompanied by a workbook and CD-ROM package that makes it ideal for teaching purposes.*
▶ *Bader, M. J., Grant, J. R., Forbes, G. S., Waters, A. J. and Lilley, R. B. E. (1997)* Images in Weather Forecasting, *Cambridge University Press. A primer on the interpretation and use of weather satellite and radar pictures. Much used for training of professional weather forecasters.*
▶ *Wallace, J. M. and Hobbs, P. V. (2006)* Atmospheric Science – An Introductory Survey, *Second edition, Academic Press. A comprehensively updated version of a meteorological classic, the first edition of which was continuously in print*

since 1977. This is another text that takes the reader into more scientific depth, pitched at the university undergraduate science student level.

WEATHER FOR SPECIFIC ACTIVITIES

▶ *Houghton, D. (1998)* Weather at Sea, *Fourth edition, Fernhurst Books. This is the recommended text on weather for sailing for the Yachtmaster's course of the UK Royal Yachting Association.*
▶ *Pedgeley, D. (2006)* Mountain Weather, *Cicerone. A detailed guide to weather in the mountains of the UK, and how avoid the worst bits of it.*

Weather websites

The nature of the internet means that the content and even existence of websites is subject to constant change. The addresses listed below have been fairly consistent over the last few years, but there's no guarantee that they will remain so in future.

NATIONAL WEATHER SERVICES

Probably the first port of call for weather information in a particular country is the website of that country's national weather service. Quality varies greatly between countries however, and while some have absolutely excellent websites, others provide only very basic information.

▶ **Australia: Bureau of Meteorology – bom.gov.au** *A wealth of weather and climate information for Australia is available on this site. There's a large section of educational material as well as all the forecasts, observations and historical climate data.*
▶ **Canada: Environment Canada – weatheroffice.gc.ca** *A very nicely designed site containing all the current weather and forecast information you need for Canada. While it contains*

less background and educational material than its US counterpart, the site is easy to navigate and good to look at.

▶ **UK: Met Office – <u>metoffice.gov.uk</u>** *The Met Office has an excellent site containing a huge amount of current, past and forecast weather information, mainly for the UK, but with forecast information for the whole world. There's also much more on this site, including information about the research that the Met Office does and educational material.*

▶ **USA: National Centers for Environmental Prediction – <u>ncep.noaa.gov</u>** *This is another site with a huge array of information. There's so much depth to this site that it seems bottomless. As well as comprehensive current and forecast information for the US there is a huge amount of educational information on topics such as El Niño, storm tracks and monsoons.*

▶ **<u>worldweather.org</u>** *The website of the United Nations World Meteorological Organization (WMO). This site has a link to the National Weather Service, if one exists, of every country in the UN. It also includes observations and city forecasts for many of these countries, provided by these National Weather Services.*

NATIONAL METEOROLOGICAL SOCIETIES

National meteorological societies are the professional bodies for meteorology in their country, but many also allow membership to students, amateur meteorologists and weather enthusiasts.

▶ **UK: Royal Meteorological Society (RMetS) – <u>rmets.org</u>** *The site contains details of the society's many activities and meetings, educational material, and a small selection of current weather information.*

▶ **USA: American Meteorological Society (AMS) – <u>ametsoc.org</u>** *The site contains details of all the society's activities. It also has a useful careers section with general information about what a career in meteorology involves, together with information on vacancies in many meteorological organizations.*

OTHER WEBSITES

A number of universities, businesses and individuals maintain excellent weather-related websites. These are perhaps more likely to be subject to changes or disappearance than those of the National Weather Services. Since some of these sites belong to commercial organizations, there may be terms and conditions attached to their use.

- ▶ **cloudappreciationsociety.org** *If you like watching clouds you'll love this fun site. Devoted to fans of clouds it includes a huge gallery of images, including a section on clouds that look like other objects (a pig with six legs for instance!).*
- ▶ **infomet.fcr.es** *A Spanish language website from the University of Barcelona. Contains a regularly updated selection of current weather data for Europe, together with forecast charts from several European Meteorological agencies.*
- ▶ **pmel.noaa.gov** *This is the website of NOAA's Pacific Marine Environment Laboratory (PMEL). If you're interested in the current state of the Pacific Ocean and particularly El Niño then this is a key website. Data from the array of moored buoys across the Pacific is displayed here in real-time, allowing you to follow the development of El Niño, not just at the surface but also beneath it.*
- ▶ **sat.dundee.ac.uk** *The website of Dundee University's satellite receiving station. Registration is required to access the satellite images provided by this site, but this is a free service available to anyone. The current images from all of the current operational geostationary satellites are available (giving global coverage), together with an archive of the past 30 days. Images from polar-orbiting satellites over Europe and the North Atlantic are also provided, including some very nice high resolution images of the UK and Europe.*
- ▶ **weather.unisys.com** *A comprehensive weather site for the USA and Canada. There are links to satellite and rainfall radar images, surface observations and NWP forecast data. An impressive part of this site is the hurricane data section. This includes information on positions, tracks and forecasts*

for all current tropical storms around the globe, together with a comprehensive archive of this information going back to 1851 in the case of Atlantic storms, and the 1940s for the other ocean basins.

▶ **weatheronline.co.uk** *Weatheronline is a commercial organization, but it provides a large amount of current, past and forecast weather information for the UK and various other European countries, as well as educational material, free of charge. They also provide interesting summaries of past weather and particularly notable weather events.*

▶ **wetterzentrale.de** *A German website with a huge selection of current weather and forecast data for Europe, including forecast data from most of the major meteorological agencies in the world. Includes a very useful archive of past temperature observations for Germany since 1876, and surface pressure charts for Europe and the North Atlantic since 1948.*

Climate change related sites

If you want to learn more about this important subject, many of the sites of National Meteorological Services listed above have sections on climate change. However, if you really want to get into the subject in detail, the best site is that of the **UN Intergovernmental Panel on Climate Change (IPCC) at ipcc.ch.** All the reports issued by this body are available on this site, together with presentations on many aspects of climate change and press releases.

Another useful site – more of a discussion group/blog – is **RealClimate.org**. This site is run by climate research scientists with contributions from a number of people working on the science of climate change. It's a good place to look to get a scientist's perspective on many of the climate change related news stories that appear in the media.

Appendix

Meteorological units

Meteorologists use a confusing array of different units to measure weather variables. Some of these units are metric, some are Imperial and some are very odd indeed, reflecting the historical development of the subject. Here is a guide to the different units that meteorologists commonly use.

ATMOSPHERIC PRESSURE

Millibar (mb) – The unit of pressure most often used in weather forecasts. Standard atmospheric pressure at sea level is usually given as 1,013 mb. An intense depression or tropical storm may have a central pressure of around 910 mb.

Pascal (Pa) – The S.I. unit of pressure, but somewhat inconvenient as a measure of atmospheric pressure and rarely used in weather forecasts and observations. 1 mb = 100 Pa.

Inches of mercury – A relic from the days when all barometers consisted of a column of mercury, the height of which was balanced by the weight of the atmospheric column above the instrument. This is not really an ideal unit for measuring atmospheric pressure since it needs to be measured to two decimal places to be meaningful. A very high surface pressure of 1,050 mb is equal to 31.01 inches of mercury and a very low surface pressure of 910 mb is equal to 26.87 inches of mercury. To convert from inches of mercury to millibars, multiply by 33.86.

Millimetres of mercury – Some barometers are calibrated in the metric equivalent of inches of mercury. 1 inch = 25.4 mm so 1,013 mb = 29.91 inches of mercury = 760 mm of mercury.

CLOUD HEIGHT AND VISIBILITY

In a standard surface observation from a weather station, the height of any clouds above the ground will be reported in feet (1 foot = 0.3 metres), whereas visibility will be reported in metres. In meteorological terms, fog is defined as a visibility less than 1,000 metres.

HUMIDITY

Specific humidity – The amount of water vapour in the air, in terms of mass of water vapour per unit mass of air. Usually expressed in grams (of water) per kilogram (of air). This unit is rarely used in weather forecasts as it doesn't mean much to the general public.

Relative humidity – A measure of the amount of water vapour in the air, expressed as a percentage of the saturated value. High relative humidity, combined with high temperature leads to uncomfortable, sticky conditions.

TEMPERATURE

Degrees Celsius or Centigrade (°C) – A temperature scale with fixed points at the freezing point and boiling point of water; 0°C and 100°C respectively.

Fahrenheit (F) – A temperature scale with fixed points at the temperature of a mixture of melting ice and salt and human body temperature; 0 F and 96 F respectively.

Kelvin (K) – The absolute temperature scale, with 0 K corresponding to absolute zero and 273 K corresponding to the freezing point of water. One Kelvin degree is the same magnitude as one Celsius degree.

To convert from Celsius to Fahrenheit: multiply by 1.8 and add 32.

To convert from Fahrenheit to Celsius: subtract 32 and divide by 1.8.

WIND SPEED

Knot – 1 nautical mile per hour. This is still the standard unit for measuring wind speed in observations from weather stations. It is a hangover from the days of sailing ships, whose speed was measured in knots. 1 nautical mile corresponds to 1/60th of a degree of latitude on the Earth's surface, which is about 1.85 kilometres.

Mile per hour (mph) – 1 mile is approximately 1.6 kilometres.

Metre per second (m/s) – This is the S.I. unit of speed, but is rarely used by weather forecasters as most people think of speeds in terms of miles or kilometres per hour. 1 m/s is approximately 2 knots or 2.2 mph.

Glossary

Advection fog – Fog which forms as warm, moist air is cooled by moving over a colder surface.

Aerosol – Microscopic solid particles, held in suspension in the atmosphere.

African Easterly Wave – An atmospheric wave that forms on the temperature gradient between the cooler surface of the Gulf of Guinea and the hot surface of the Sahara. The wave moves westwards and can trigger Atlantic hurricane formation.

Air mass – A large expanse of the atmosphere through which the air has acquired very uniform characteristics, usually due to the presence of a stationary anticyclone.

Albedo – A measure of the ability of a surface to reflect the energy coming from the Sun. Snow, ice and clouds have a high albedo, whereas dark surfaces like bare soil have a low albedo and so will absorb most of the incident solar radiation.

Altocumulus – A medium-level cloud with a wide variety of different forms (see colour plate section for some examples).

Altostratus – A medium-level cloud with a layered, rather formless appearance.

Ana-front – An active frontal system with vigorous vertical motion leading to thick cloud and extensive precipitation.

Anticyclone – A region of relatively high atmospheric pressure, associated with settled weather conditions. Winds blow in a clockwise direction around an anticyclone in the northern hemisphere and in an anticlockwise direction in the southern hemisphere. Winds near the centre of an anticyclone are light or calm.

....

Atmospheric pressure – Weight of a column of air above a point. The pressure is measured by a barometer.

....

Background field – The state of the atmosphere used to provide the starting point for setting the initial conditions for a computer-generated weather forecast. The background field is usually a short-range forecast from the previous run of the computer forecast.

....

Backing – An 'anticlockwise' change in direction from which the wind is blowing. A westerly wind which changes to become south-westerly with time would be described as backing.

....

Baroclinic zone – Region of strong horizontal temperature gradient. A preferred location for the development of mid-latitude depressions.

....

Bogussing – Insertion of phoney observations into the data assimilation cycle of an NWP model in order to force a particular development.

....

Bora – Cold downslope wind which blows from the Balkans down the Adriatic Coast.

....

Chinook – Warm, downslope wind which blows down the eastern side of the Rockies, raising the temperature rapidly and often resulting in snowmelt.

....

Cirrocumulus – Form of high cloud made from ice crystals, often having a rippled or speckled appearance.

....

Cirrostratus – Form of high ice cloud, often appearing as a milky white sheet.

Cirrus – High ice cloud with the appearance of fibrous tufts or filaments.

Climate change – Variations in average weather conditions, usually over decadal or longer timescales. Can be natural or human induced (see *global warming*).

Cloud condensation nuclei (CCN) – Microscopic solid particles in suspension in the atmosphere, which act as sites for the condensation of water in clouds.

Convection – Vertical motion of air due to buoyancy forces, usually caused by heating of air parcels in contact with a warm surface. Can lead to the formation of cumulus-type clouds.

Convergence – Accumulation of air at a location due to the horizontal winds. Convergence near the Earth's surface can lead to ascending vertical motion.

Conveyor belt – Flow of air through or around a mid-latitude depression.

Coriolis effect – The Earth's rotation causes parcels of air in motion to be deflected from their original path, to the right in the northern hemisphere and to the left in the southern hemisphere.

Cumulus – Cloud with a lumpy or heaped-up appearance, formed as the result of convection.

Cumulonimbus – Type of cumulus cloud with extensive vertical development, often containing ice in the upper parts. Produces precipitation and possibly hail and thunder.

Cut-off low – Depression which forms as the result of cold air from polar latitudes being stranded at lower latitudes due to a meander in the jet stream.

Cyclone – Generic term for any type of low-pressure weather system. Often applied to tropical storms in the Indian Ocean.

Data assimilation – Blending of NWP model background field and weather observations to set the initial conditions for a computer weather forecast.

Depression – Another generic term for any low-pressure weather system.

Development area – Region of the atmosphere where depressions often form or deepen. Usually associated with features in upper-level jet streams.

Dewpoint – Temperature to which air would need to be cooled to cause condensation of water vapour.

Divergence – The opposite of convergence. Air is removed from a region more rapidly than it is replaced by the horizontal winds. Upper-level divergence often leads to ascent of air through the troposphere.

Downdraught – Downwards current of air in a convective cloud, driven by cooling due to evaporation of raindrops.

Electromagnetic radiation – Form of energy which can travel through a vacuum. All the energy that reaches the Earth from the Sun is in the form of electromagnetic radiation. The Earth also emits electromagnetic radiation, mostly in the infrared part of the spectrum.

El Niño – Periodic warming of the equatorial Pacific Ocean, occurring every 3–7 years.

Emissions scenario – Prediction of the amount of greenhouse gases that are likely to be emitted into the atmosphere by human activity.

Ensemble forecasting – Running a weather forecast many times with different initial conditions to assess the predictability of the atmosphere and generate probability forecasts.

ENSO (El Niño – Southern Oscillation) – Coupled ocean-atmosphere variability in the tropical Pacific. Warming of the ocean surface in the centre and east of the basin leads to changes in the circulation patterns, cloud and precipitation across the whole basin and further afield.

Extra-tropics – Those parts of the globe outside the tropical regions, i.e. polewards of latitudes of about 23 degrees north and south (see also *tropics*, *tropical regions*, *subtropics*).

Föhn – Warm downslope wind that blows in the Alps. Often used to describe any warm downslope wind in other mountain areas.

Front – Boundary between two differing air masses, often associated with a band of thick cloud and rain.

Geostationary satellite – Satellite placed in orbit at such an altitude that its orbit period is 24 hours, and so it remains fixed over a point on the Earth's surface.

Geostrophic wind – Theoretical wind that would result from an exact balance between the pressure gradient force and the Coriolis force. Can be determined from the spacing of isobars on a weather map.

Global warming – Increasing global average temperature as a result of human activity increasing the levels of greenhouse gases in the atmosphere.

Gradient – Rate of change of a particular meteorological quantity with distance.

Greenhouse gas – Any atmospheric gas that absorbs the radiation emitted by the Earth's surface, thus warming the atmosphere. Carbon dioxide, water vapour, ozone and methane are all greenhouse gases.

Gust front – Sudden increase in surface wind, usually associated with a convective downdraught hitting the ground and spreading out horizontally.

Hadley cell – Circulation of air in the vertical plane in the tropics, with a rising branch at the ITCZ, sinking over the subtropics and surface trade winds blowing towards the equator.

Haze – Reduction in visibility due to the presence of microscopic solid particles in the atmosphere, either due to man-made pollution or natural sources such as wind-blown dust.

Hindcast – NWP forecast made after the event to test the ability of the model to predict a particular weather situation.

Hurricane – A tropical cyclonic weather system in the Atlantic sector, with mean winds in excess of 119 km/h (74 mph).

Ice storm – Rain which falls onto a sub-zero surface, freezing on contact and thus coating the surface with a layer of clear ice, sometimes many millimetres thick.

Inter-tropical convergence zone (ITCZ) – Region of the tropics at which the trade winds from the two hemispheres meet. Characterized by light winds, deep convection and heavy rain.

Isobar – Line on a weather map joining points of equal pressure.

Jet stream – Narrow belt of strong winds in the upper part of the troposphere.

Jet streak – Small zone of excessively strong winds embedded within the jet stream.

Kata-front – Weak, inactive weather front, with little or no precipitation.

Latitude – Position on the Earth's surface relative to the equator. The equator is defined as having a latitude of 0 degrees, with the North Pole and South Pole having latitudes of 90 degrees north and 90 degrees south respectively. London is at a latitude of about 51 degrees north, and New York is at a latitude of about 40 degrees north.

Lake effect snow – Heavy snow on the lee-side of a large body of unfrozen water. Caused by cold air picking up moisture as it flows across the lake and then dumping it out as snow on the far side.

Land breeze – Offshore wind formed due to cooling of the land relative to the adjacent ocean. Usually occurs at night.

La Niña – Periodic cooling of the eastern equatorial Pacific Ocean, often following an El Niño event.

Lenticular cloud – Lens- or saucer-shaped cloud resulting from airflow over a hill in stable conditions.

Longitude – Position of the Earth's surface relative to the Greenwich Meridian which passes through London. London thus has a longitude of 0 degrees, and the International Date Line in the Pacific Ocean largely coincides with a longitude of 180 degrees. New York has a longitude of 74 degrees west and Delhi has a longitude of 77 degrees east.

Madden-Julian Oscillation (MJO) – Variation between large-scale cloudy, wet conditions and clear, dry conditions in the equatorial Indian and Pacific Oceans, on a 30–90 day period.

Mesosphere – Layer of the atmosphere lying between about 50 kilometres and 85 kilometres above the Earth's surface.

Mid-latitude – That part of the globe lying between about latitudes of about 40 and 70 degrees in each hemisphere. The weather here is characterized by the relative dominance of mid-latitude depressions and anticyclones.

Mistral – Strong wind caused by funnelling of air down the Rhône Valley in southern France.

Monsoon – Climate characterized by distinct wet and dry seasons, usually accompanied by large changes in wind direction.

Nimbostratus – Thick, layered, rain-bearing cloud often associated with fronts.

Nowcasting – Continuous real-time monitoring of local weather conditions, and issuing of short-range forecasts for the next few hours. Weather radar is an important tool in this process.

Numerical Weather Prediction (NWP) – Use of computer models of the atmosphere, based on the laws of physics, to forecast the weather. Requires powerful computers.

Ozone – Minor atmospheric trace gas that absorbs ultraviolet wavelengths of radiation from the Sun, thus shielding the surface from this harmful radiation. Also acts as a greenhouse gas.

Ozone layer – That part of the atmosphere, between about 20 kilometres and 30 kilometres above the surface, that contains the majority of the ozone in the atmosphere.

Polar cell – Weak vertical circulation at high latitudes, with rising motion at around 70 degrees and sinking motion over the poles.

Polar low – Small depression that forms over the oceans at high latitudes as cold air is heated from below by movement over a warmer ocean surface.

Polar orbiting satellite – Weather satellite in an orbit at an altitude of about 85 kilometres. Orbits over the poles and passes over each point on the Earth's surface twice a day.

Radiation see *Electromagnetic radiation*.

Radiation fog – Fog which forms over land on cold, clear nights as the surface is cooled by radiation of energy to space. Air in contact with the surface is cooled to its dewpoint and condensation occurs.

Radiosonde – A package of instruments for measuring temperature, humidity and pressure, attached to a weather balloon.

Rainfall radar – Form of radar tuned to measure rainfall intensity.

Relative humidity – Measure of the water vapour content of a sample of air, compared to its maximum possible value at saturation.

Sea breeze – Onshore wind that blows as the result of daytime heating of land relative to the adjacent ocean surface.

Seeder-feeder mechanism – Enhancement of rainfall rates from frontal cloud (feeder cloud) as the rain falls through a lower layer of stratus (seeder cloud) caused by airflow over hills.

Short wave trough – Small change in direction of an upper-level jet stream.

Smog – Strong reduction in visibility due to a combination of large amounts of pollutants in the lower atmosphere and high humidity.

Solar constant – The amount of energy from the Sun which reaches the top of the Earth's atmosphere. Usually around 1,370 watts per square metre.

Specific humidity – A measure of water vapour content of a sample of air, expressed as grams of water vapour per kilogram of air.

Split cold front – Boundary between a warm and a cold air mass in which cold, dry air is overrunning the surface front, thus moving the upper front ahead of the surface feature.

Stability – Measure of the likelihood of convective activity in the atmosphere. Convection cannot occur in a stable atmosphere.

Storm surge – Rapid rise in sea level at a coast associated with strong along-shore winds.

Storm track – The average path taken by depression systems in a particular sector of the globe.

Stratocumulus – Layered cloud made of water droplets, with some appearance of lumpiness within the layer.

Stratosphere – Layer of the atmosphere between about ten kilometres and 50 kilometres above the Earth's surface. Characterized by temperature increasing with height due to the presence of the ozone layer.

Stratus – Very low, layer cloud either in the form of an extensive sheet or broken, ragged clouds.

Subtropics – Those parts of the globe between about 20 and 40 degrees of latitude north and south of the equator. The weather in these regions is characterized by high atmospheric pressure, largely clear skies or thin cloud, and little precipitation.

Surface analysis – Weather map showing the positions of fronts, depressions, anticyclones and isobars at any given time.

Temperature inversion – Region of the atmosphere in which temperature is increasing with height.

Thermohaline circulation – Global circulation of water in the oceans, driven by differences in temperature and salinity of water in different ocean basins.

Thermocline – Large vertical temperature gradient in the oceans, separating warm surface water and colder, deep water layers.

Thermosphere – Upper layer of the atmosphere, extending upwards of about 85 kilometres. Air is very thin in this layer and temperature variations can be very large.

Topography – Hills, mountains, valleys, etc.

Trace gases – Many atmospheric gases which between them make up less than one per cent of the total atmospheric mass.

Trade winds – Consistent easterly winds blowing across the tropical oceans, with a marked component blowing toward the equator.

Tropics, tropical region – That region of the globe between latitudes of about 23 degrees north and 23 degrees south. Characterized by strong heating from the Sun and weather systems dominated by convection (see also *subtropics*, *extra-tropics*).

Troposphere – Lowest layer of the Earth's atmosphere, extending from the surface up to about 10–15 kilometres. Temperature generally falls with altitude in the troposphere, which also contains most of the water vapour within the atmosphere.

Tropopause – Boundary between the troposphere and the stratosphere. Characterized by a layer where temperature remains constant with height, and then starts increasing.

Typhoon – Strong tropical depression in the West Pacific.

Updraught – Rising convective current of warm air.

Urban heat island – Tendency of urban areas to be warmer than surrounding countryside, particularly at night. This is due to the concrete and tarmac heating up strongly during the day and then releasing that heat back to the surrounding air during the night.

Veering – 'Clockwise' change in wind direction over time. A westerly wind that moves round to become north-westerly is veering.

Walker circulation – Circulation of air in a vertical plane along the equator. In the Pacific the Walker circulation is characterized by rising air in the west, upper-level westerly winds, sinking in the east and surface easterly trade winds blowing back towards the west.

Warm sector – Region of warm, moist air of tropical origin on the equatorward side of a mid-latitude depression. Bounded by the warm and cold fronts.

Weather system – Any region of the atmosphere in which the weather takes on particular characteristics distinct from the conditions surrounding that region.

Windshear – Change in direction and/or speed of the wind with height.

Index